NASA STI Program . . . in Profile

Since its founding, NASA has been dedicated to the advancement of aeronautics and space science. The NASA Scientific and Technical Information (STI) program plays a key part in helping NASA maintain this important role.

The NASA STI Program operates under the auspices of the Agency Chief Information Officer. It collects, organizes, provides for archiving, and disseminates NASA's STI. The NASA STI program provides access to the NASA Aeronautics and Space Database and its public interface, the NASA Technical Reports Server, thus providing one of the largest collections of aeronautical and space science STI in the world. Results are published in both non-NASA channels and by NASA in the NASA STI Report Series, which includes the following report types:

- TECHNICAL PUBLICATION. Reports of completed research or a major significant phase of research that present the results of NASA programs and include extensive data or theoretical analysis. Includes compilations of significant scientific and technical data and information deemed to be of continuing reference value. NASA counterpart of peer-reviewed formal professional papers but has less stringent limitations on manuscript length and extent of graphic presentations.

- TECHNICAL MEMORANDUM. Scientific and technical findings that are preliminary or of specialized interest, e.g., quick release reports, working papers, and bibliographies that contain minimal annotation. Does not contain extensive analysis.

- CONTRACTOR REPORT. Scientific and technical findings by NASA-sponsored contractors and grantees.

- CONFERENCE PUBLICATION. Collected papers from scientific and technical conferences, symposia, seminars, or other meetings sponsored or cosponsored by NASA.

- SPECIAL PUBLICATION. Scientific, technical, or historical information from NASA programs, projects, and missions, often concerned with subjects having substantial public interest.

- TECHNICAL TRANSLATION. English-language translations of foreign scientific and technical material pertinent to NASA's mission.

Specialized services also include creating custom thesauri, building customized databases, organizing and publishing research results.

For more information about the NASA STI program, see the following:

- Access the NASA STI program home page at *http://www.sti.nasa.gov*

- E-mail your question via the Internet to *help@sti.nasa.gov*

- Fax your question to the NASA STI Help Desk at 301–621–0134

- Telephone the NASA STI Help Desk at 301–621–0390

- Write to:
  NASA Center for AeroSpace Information (CASI)
  7115 Standard Drive
  Hanover, MD 21076–1320

NASA/CR—2008-215440/PART2

Hydrogen Research for Spaceport and Space-Based Applications
Hydrogen Sensors and Systems

Tim Anderson and Canan Balaban
University of Florida, Gainesville, Florida

Prepared under Grant NAG3–2930

National Aeronautics and
Space Administration

Glenn Research Center
Cleveland, Ohio 44135

November 2008

Trade names and trademarks are used in this report for identification
only. Their usage does not constitute an official endorsement,
either expressed or implied, by the National Aeronautics and
Space Administration.

This work was sponsored by the Fundamental Aeronautics Program
at the NASA Glenn Research Center.

Level of Review: This material has been technically reviewed by NASA technical management.

Available from

NASA Center for Aerospace Information 7115 Standard Drive Hanover, MD 21076–1320	National Technical Information Service 5285 Port Royal Road Springfield, VA 22161

Contents

1. Integration and Testing of Low Power Wireless Hydrogen Sensor Modules 3
2. Novel ZnO Nanorod Hydrogen Gas Sensors 27
3. Modeling of ZnO Nanorod Hydrogen Gas Sensors 41
4. Environmentally - Driven Power Source, Power for Wireless Hydrogen Sensor Network - Energy Harvesters 51
5. Detecting Hydrogen by Enzyme-Catalyzed Electrochemistry 65
6. Micro-machined Floating Element Hydrogen Flow Rate Sensor 83
7. Ultra-wideband Communication for Tiny Low Power Radios 89

Hydrogen Research for Spaceport and Space-Based Applications

Hydrogen Sensors and Systems

Tim Anderson and Canan Balaban
University of Florida
Gainesville, Florida 32611

Abstract

The activities presented are a broad based approach to advancing key hydrogen related technologies in areas such as fuel cells, hydrogen production, and distributed sensors for hydrogen-leak detection, laser instrumentation for hydrogen-leak detection, and cryogenic transport and storage. Presented are the results from research projects, education and outreach activities, system and trade studies. The work will aid in advancing the state-of-the-art for several critical technologies related to the implementation of a hydrogen infrastructure. Activities conducted are relevant to a number of propulsion and power systems for terrestrial, aeronautics and aerospace applications.

Sensor systems research was focused on hydrogen leak detection and smart sensors with adaptive feedback control for fuel cells. The goal was to integrate multifunction smart sensors, low-power high-efficiency wireless circuits, energy harvesting devices, and power management circuits in one module. Activities were focused on testing and demonstrating sensors in a realistic environment while also bringing them closer to production and commercial viability for eventual use in the actual operating environment.

1. Integration and Testing of Low Power Wireless Hydrogen Sensor Modules

Task PI: Dr. Jenshan Lin, Electrical & Computer Engineering, University of Florida

Collaborators: Dr. F. Ren, Chemical Engineering, Dr. D. Norton, Dr. S. Pearton, Material Science & Engineering, Dr. K. Ngo, and Dr. T. Nishida, Electrical and Computer Engineering, University of Florida

Research Period: August 3, 2004 to March 31, 2008

Abstract

A self-powered wireless hydrogen sensor node has been designed and developed from a system level approach. By using multi-source energy harvesting circuitry such as scavenged or "reclaimed" energy from light emitting and vibrational sources as the source of power for commercial low power microcontrollers, amplifiers, and RF transmitters, the sensor node is capable of conditioning and deciphering the output of hydrogen sensitive ZnO nano-rods sensors. Upon the detection of a discernible amount of hydrogen, the system will 'wake' from an idle state to create a wireless data communication link to relay the detection of hydrogen to a central monitoring station. Two modes of operation were designed for the use of hydrogen detection. The first mode would sense for the presence of hydrogen above a set threshold, and alert a central monitoring station of the detection of significant levels of hydrogen. In the second mode of operation, actual hydrogen concentrations starting as low as 10 ppm are relayed to the receiver to track the amount of hydrogen present.

In addition, a low-power wireless hydrogen sensing system has been designed and developed for field test in a facility in Orlando. The system uses AlGaN/GaN High Electron Mobility Transistors (HEMTs) as the sensing devices. The sensors have achieved ppm level detection, with the added advantages of a very rapid response time within a couple of seconds, and rapid recovery. The sensors have shown current stability for more than 8 months in an outdoor environment. The wireless network sensing system enables wireless monitoring of independent sensor nodes that transmits wireless signals to a central monitoring station. This is especially useful in manufacturing plants and hydrogen-fuelled automobile dealerships, where a number of sensors would be required. The software installed on the central monitoring station can be programmed to trigger the alarm or transmitting messages to emergency contacts through phone network when hydrogen concentration over a preset threshold is detected. The data has also been collected remotely through internet at University of Florida for long term analysis.

Introduction

The first objective of this research task is to develop technologies to integrate novel nano-scale sensors, low-power electronic circuits, and energy harvesting devices to demonstrate the feasibility of a self-powered wireless hydrogen sensor. The self-powered sensor module will operate without the need of replacing battery. Together with reusable ZnO nanorod sensors, it ensures a very long lifetime operation without the need of calibration and replacement. The PI and his research group has been focusing on the design of low-power electronic circuits interfacing with low-power sensors to minimize the load to energy harvesting devices. The integrated sensor module provides a testing platform to test the effectiveness of different sensor devices and energy harvesting devices. The proposed self-powered wireless hydrogen sensor module reduces the size and cost of each module and the overall maintenance cost of the hydrogen leak detection system.

The second objective of this research task is to develop a complete system solution for field test by integrating the sensors with low-power electronic circuits, radio frequency transceivers, power management circuits, and software algorithms to control data transmission and processing. In addition, the system aims to provide real time monitoring and alarm capabilities via the Internet as well as the cellular phone network.

The integrated sensor module provides a platform to test the effectiveness of different sensor devices as well as energy harvesting devices. The proposed self-powered wireless hydrogen sensor module will reduce the size and cost of each module and the overall maintenance cost of the hydrogen leak detection system.

Background

Hydrogen has been the primary energy source in NASA's space exploration projects. The use of hydrogen ranges from rocket propellant to fuel cell. The extensive use of hydrogen requires advanced technologies to ensure its safety in production, transport, and storage. In addition to space exploration missions, President Bush announced the Hydrogen Fuel Initiative in his January 2003 State of the Union Address. The Hydrogen Fuel Cell Initiative will push the hydrogen technologies for civilian applications (industrial, residential, transportation, and utility) to develop clean energy source powering our vehicles. With increased interest in hydrogen-powered energy system, it is expected that a substantial increase of research and development effort is needed to support this energy revolution. To gain public acceptance of this transition to a hydrogen-powered energy system, safety is the first issue to deal with. A catastrophic failure in any hydrogen project could irreparably damage the entire transition strategy. The reliable and accurate detection of hydrogen leakage during its production, storage, distribution, and use is therefore very important. It is crucial to have a robust hydrogen sensor system to accurately detect any hydrogen leakage in its early stage before it reaches a level that may cause catastrophe.

The hydrogen sensors can also be used to monitor the hydrogen concentration in various hydrogen systems including process plants, transport systems, storage systems, and fuel cells. The accurate reading of hydrogen concentration enables the hydrogen energy system to have better control over the hydrogen production, delivery, and the amount of energy produced.

Specific NASA mission and technical needs to be addressed
NASA's space exploration missions require extensive use of hydrogen. Hydrogen first needs to be produced and stored safely. Hydrogen will then need to be transported to the mission launch center, e.g., Kennedy Space Center, right before the mission. From the beginning when hydrogen is produced to the end when it is used, hydrogen has to be handled with extreme care during its production, storage, distribution, and use. A hydrogen sensor system is needed to detect tiny leakage and sends the signal to a central monitoring station for immediate attention. Since the leakage can occur at any possible location that handles the hydrogen, the effective detection requires an enormous number of sensors. To instrument large numbers of distributed hydrogen sensors, the sensors need to operate autonomously without wires. Hence, the sensor output must be transmitted wirelessly. Furthermore, the power to operate the sensors must be self-contained. Since batteries have a finite lifetime, and it is impractical to replace the batteries

for an enormous number of sensors, a self-powered wireless hydrogen sensor system is necessary to safely monitor the hydrogen resource network. These sensors and the monitoring station have to work well together in a system. The hydrogen sensor system should be designed for easy deployment. Small size, robustness, and low cost are key factors. The sensor system should be scalable to cover from small to large area.

The hydrogen sensor system is mainly used for safety assurance. When integrated into a storage system or a delivery piping system, the sensor system needs to indicate hydrogen permeation through a composite tank, leaks within a closed storage space, or can be used to determine if fuel delivery piping or storage tank has sustained critical damage which could result in a catastrophic failure. Innovative concepts and devices are needed to develop a sensitive and robust hydrogen sensor system to achieve this requirement. These concepts or devices shall be used to control the safety systems in vehicles and stationary appliances and must prevent the buildup of hydrogen, e.g., actuate valve closures or ventilation devices.

For pipeline hydrogen delivery system, we need an innovative in-situ sensing technology for pipeline leaks and material failures that allows the development and adoption of standards for the direct burial of the pipeline, but does not require exposure of the sensor to verify leaks or a system failure. A built-in wireless connection is well suited for this application. The sensor data can be transmitted by radio waves to a monitoring station.

In process plants, hydrogen will be produced from a number of feedstock and at many different scales. The ability to detect leaks before personnel enter an area or measure rising levels of hydrogen while working in an area are essential to the safety in process plants. Therefore, sensing technologies for both large area and personnel safety are needed.

The hydrogen sensor system is also needed to monitor hydrogen concentration in many applications. The sensors need to provide accurate readings of hydrogen concentration from low level to high level and with fast respond time. The sensor should be designed for repeated use, not disposable. Such a system will be very useful to monitor the production, transport, and storage of hydrogen, and can be used in experiments to verify the model of hydrogen transport system. In addition, hydrogen sensors and sensors for flow, temperature, pressure, and humidity are needed for efficient operation and control of proton exchange membrane (PEM) fuel cells, solid oxide fuel cells (SOFC), and aircraft fuel cell reformers.

The specific targets of the hydrogen sensor system are listed in the table below.

Feature	Requirement
Small size and form factor	The sensor device including the wireless transceiver circuit should be small and low profile that can be easily integrated into hydrogen systems.
Robustness and reliability	The sensor should be able to operate at a wide range of environment conditions. The typical temperature range is –50 °C to 120 °C.
Sensitivity	The sensor should be able to detect hydrogen concentration level as low as 1 ppm.
Response time	Response time should be less than 1 sec when hydrogen concentration increases to 10%.

Feature	Requirement
Reusability	The sensor should be designed for repeated use within its lifetime. The recovery time of sensor from high levels of hydrogen concentration to normal ambient air condition should be no more than 30 sec.
Very long lifetime	The sensor should have lifetime longer than 10 years. Within the lifetime there should be no need to maintain the remote sensors, including calibration, inspection, replacing batteries, etc.
Wireless transmission of sensor data	The sensor data should be transmitted to a remote central monitoring station wirelessly.
Self-powered operation	The sensor should be self-powered to avoid the need of replacing batteries.

To meet the above requirements, several key technical areas need to be addressed. The sensor system development requires interdisciplinary research collaboration from material, device, circuit, to system. This includes novel materials and devices, advanced analog and mixed signal circuits, and wireless communication system. The mission of this research task is to lead the interdisciplinary research collaboration to create the integrated system solution.

Experimental

There is currently great interest in the development of hydrogen sensors for applications involving leak detection in hydrogen fuel storage systems and fuel cells for space craft [1]-[4]. One of the most important aspects desired by the end user for such a sensor is the ability to selectively detect hydrogen at room temperature with the presence of air in the ambient. In addition, for most of these applications, the sensors are also required to have very low power consumption and minimal weight. Due to the intrinsic characteristics of nanostructures, they naturally become formidable candidates for this type of sensing. Since the power requirement for operating the sensor is extremely small, the sensor and a leakage detecting transmitting system can be run off of energy harvesting devices.

ZnO based nano-rods have many unique characteristics which make them fundamentally appropriate candidates for the sensing of hydrogen. ZnO is a material currently used in the detection of pH, humidity, UV light, and gas, and has shown to change resistance with respect to both temperature and hydrogen exposure [5]-[8]. Because of its wide bandgap of 3.2eV, the ease of synthesizing nanostructures, the availability of heterostructures, and the bio-safe characteristics of this material, ZnO is a most attractive material for the specific sensing application at hand. With ZnO nano-rods placed in an array, as a gas sensor, they are able to create a larger chemically sensitive surface-to-volume ratio which is needed for high sensitivity in hydrogen sensing. ZnO nano-rods can also be produced cheaply, and are highly compatible with other microelectronic devices [9]-[11].

In this research task, we conduct the experiment on a ZnO nano-rods based hydrogen sensing system which consists of nano-rod sensor, solar and vibrational energy harvesting devices with power management circuits, a sensor interface, microcontroller, and the RF front end of the sensor system. The design and optimization of the detection circuitry, digital processing considerations, and modulation scheme to maintain an accurate and reliable system with a minimal amount of energy scavenged will be described in the next section.

We also conducted field test of a low-power wireless hydrogen sensor network in Ford Greenway at Orlando. The facility provides regular maintenance service to several hydrogen-powered vehicles. The field test result would provide useful data for analysis.

Results and Discussion

Integration and Testing of Self-Powered Wireless Hydrogen Sensor Modules

Fig. 1 shows the transient response to different dopings of hydrogen for ZnO nano-rods coated with Pd. Note that no currents were measured through the discontinuous Au islands and no thin film of ZnO on the sapphire substrate was observed with the growth condition for the nanorods. Therefore the measured currents were due to transport through the nanorods themselves. The I-V characteristics from the multiple nanorods were linear with typical currents of 0.8 mA at an applied bias of 0.5 V.

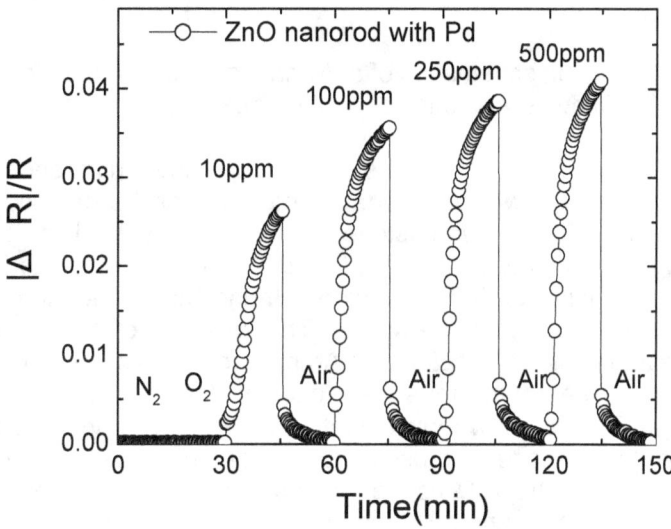

Fig. 1: Transient response of Pd-coated ZnO Nano-Rods for different hydrogen dopings

The challenge in designing the interface between a sensor and the Analog-to-Digital (A/D) converter of a system was to obtain an accurate real world signal with the limitations of low power and reduced voltage swings. Given that the ZnO nano-rod's initial response to any exposure of hydrogen was distinct and immediate, this intrinsic characteristic served as an ally for the successful detection of hydrogen. Since the ZnO nano-rod's resistance changed with respect to how much and how long the device has been exposed to hydrogen, the most popular and accurate way of detecting resistance changes was through the use of Wheatstone Resistive Bridges, as illustrated in Fig. 2(a). The main objective of the Wheatstone Resistive Bridge stage is to detect the differences in resistance between a ZnO nano-rod and a passivated ZnO nano-rod. The passivated ZnO nano-rod was encased with silicon nitride. By using a passivated ZnO nano-rod encased in glass to be the resistor, only the resistance changes caused by exposure to hydrogen was detected, and not from other variables such as temperature. With no hydrogen present, the passivated and exposed ZnO nano-rods were similar in resistance, and the output voltage of our sensor to A/D converter interface was approximately 0 V, which is very close to the ideal condition.

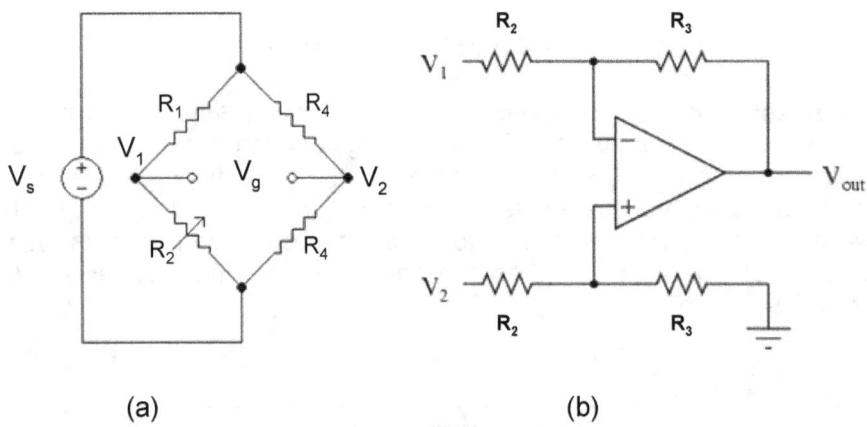

Fig. 2: (a) Circuit schematics of a Wheatstone resistive bridge
(b) Circuit schematics of a difference amplifier

By using the combination of a Wheatstone Resistive Bridge, and a difference amplifier as shown in Fig. 2(b), an extremely low power and portable interface for the detection of hydrogen was designed, as shown in Fig. 3. The Wheatstone Resistive Bridge and the Difference Amplifier were combined through an additional non-inverting gain amplifier stage to buffer, amplify, and provide a high impedance input to the Wheatstone Bridge before the signal from the resistive bridge was processed by the difference amplifier. This topology of amplifiers is also known as an instrumentation amplifier. MAX4289 Op-Amps from Maxim-IC were used as the instrumentation amplifier. These Op-Amps were chosen due to their low power requirements (1.0 V/9 uA) and for their typical low input offset voltage of 200 µV. By applying a reference voltage of 2 V on the supply voltage, and using the onboard 10- bit A/D converter of a low power microcontroller, the MSP430, the A/D converter had a resolution of about 2 mV, with 1024 digitized voltage levels between 0 and 2 V. The output of the interface met this requirement and provided at least a 2 mV per ohm (output voltage to ZnO resistance change) output. The measurements for the fabricated sensor interface are listed in Table 1. Given the A/D resolution of 2 mV, the sensor and sensor interface were capable of detecting the presence of at least 10 ppm of hydrogen.

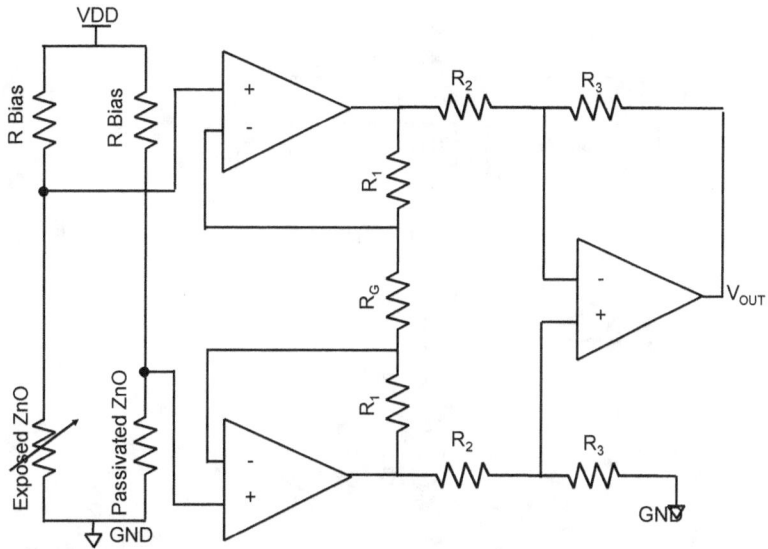

Fig. 3: Complete ZnO nano-rod to ADC differential detection interface

Table 1 Sensor Interface Measurements

Supply Voltage	ZnO Resistance	PPM	%ΔR/R change	Supply Current	Power	Sensor Output
2 V	1565	0	0	42 μA	84 μW	80 mV
2 V	1521	10	2.75	44.2 μA	88.4 μW	138 mV
2 V	1500	500	4.2	44.3 μA	88.6 μW	240 mV

The complexity of the modulation was the key component in our selection of a wireless RF front end. To satisfy the RF transmitter's requirement for ultra-low power consumption operation without sacrificing our transmission distance, a lower modulation scheme equaled lower transmitter architecture complexity was used to realize a RF front end with fewer discrete power consuming components. The simplest modulation scheme available was a "carrier present, carrier absent" technique, also known as On Off Keying (OOK). The appeal of the OOK modulation scheme is the simple premise that an OOK transmitter would only be "on" and consuming power when transmitting a "high" or a "1", and that little to no power consumed on the transmission of a "low" or "0".

For the RF front end of our sensor, a Ming TX-99 transmitter tuned to 300 MHz was used. The architecture of this transmitter based on a Colpitt Oscillator design is shown in Fig. 4. It consisted of a single transistor and a LC tank to tune the transmitter to oscillate at a specific frequency. The transmitter current consumption was 850 μA at a bias voltage of 0.6 V, which translated a power requirement of 510 μW to transmit a constant 50% duty cycle and 580 mV peak to peak square pulse train at 1 kHz. A quarter-wavelength 22 gauge copper monopole antenna tapped at one of the inductors of the LC tank was also used to increase our transmission distance.

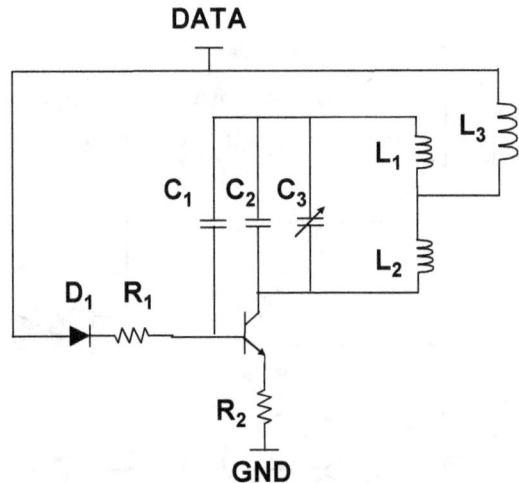

Fig. 4: Circuit schematic of the wireless transmitter

These transmitters communicated to a central monitoring station consisting of a receiver, data acquisition device, and a laptop. The receiver was a LC based Ming RE-99 receiver module tuned to 300 MHz. The data output of the RF Module was tied to a Labview USB-6008 data acquisition (DAQ) devices which deciphered the output of the receiver and displayed the information onscreen in the LabView code. Both the DAQ device and the receiver module were powered from the USB port of a laptop, so power consumption was of little worry to the development of the central monitoring station.

The proper selection of our microcontroller was essential to the success of the design. It was a requirement that the microcontroller included an onboard A/D converter with enough resolution to track the changes of the sensors, and be capable of conditioning and processing the data received from the sensor interface. Sufficient onboard memory was desired to retain both the runtime code as well as store the data sampled by the A/D converter. Most importantly, the microcontroller needed to have the ability to encode and send this data to the transmitter via a serial output. Due to the use of the OOK in the modulation scheme, the system was optimized with a serial output port capable of sourcing enough power to drive and power our transmitter included in the microcontroller. The Texas Instrument's MSP430F1232IPW was chosen as the microcontroller, which met the requirements. Table 2 highlights all the pertinent features of the MSP430F1232IPW.

Table 2 Features of TI MSP430F1232IPW

Type of Program Memory	Flash
Program Memory	8 kB
RAM	256 Bytes
I/O Pins	22 pins
ADC	10-bit SAR
Interface	1 Hardware SPI or UART, Timer UART
Supply Voltage Range	1.8 V – 3.6 V
Active Mode	200 uA @ 1 MHz, 2.2 V supply
Standby Mode	0.7 uA
# of Power Saving Modes	5

The microcontroller was programmed to run as a state machine, and had two different reprogrammable modes of operation. In each mode of operation, the microcontroller operated within the following states: initialize, collect data, transmit data, and sleep. The first mode of operation was for the level monitoring of hydrogen. This mode ran through each state until a discernable threshold of hydrogen was detected. This threshold was set at a level that hydrogen concentration would be high enough to pose serious danger. Once this level of hydrogen concentration was detected, the microcontroller forced the RF front-end to transmit an emergency pulse to the central monitoring station.

The second mode of operation was for data transmission. In this mode, the microcontroller collected data from the sensor interface, and queued this data to the RF front-end to be transmitted to the central monitoring station. This mode was for a constant tracking of hydrogen levels, while the level monitoring mode was to alert the end user that hydrogen has indeed been detected. The state flow diagrams for the Level Monitoring Mode and Data Transmission Mode are shown in Fig. 5.

The objectives in the design of our energy harvesting devices involved the ability to extract energy from two main sources of energy – solar and vibrational devices. The end product was one that was able to extract energy from both photovoltaic devices and piezoelectric (PZT) energy harvesters. This energy would then be primed for use by the hydrogen sensor, and sensor interface, as well as the micro controller and transmitter. A system block diagram of the energy harvesting devices is illustrated in Fig. 6.

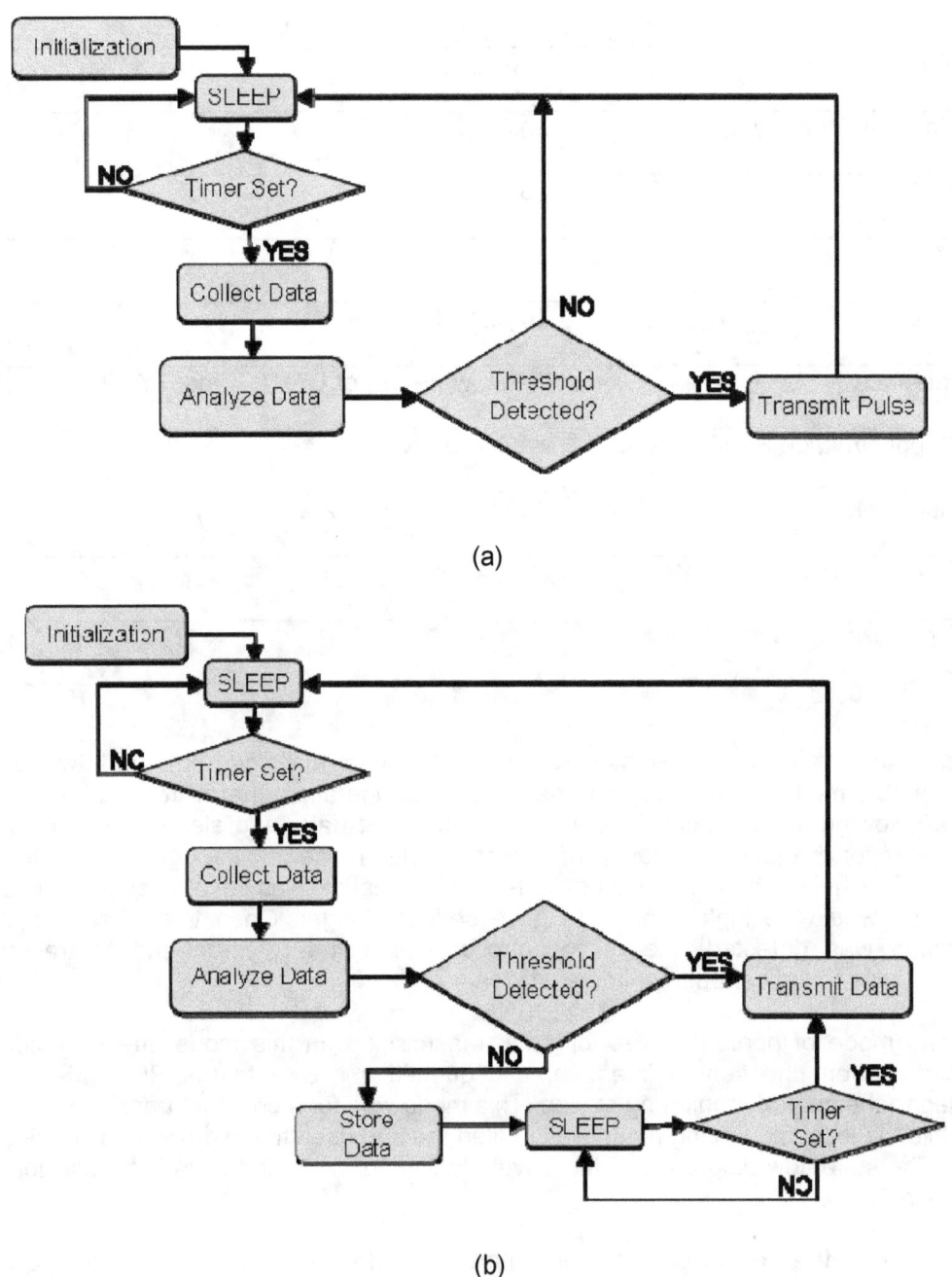

Fig. 5: (a) State flow diagram of level monitoring
(b) State flow diagram of data transmission

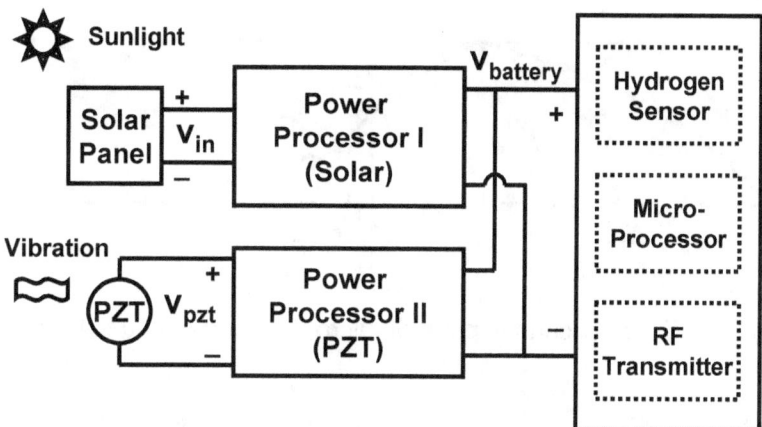

Fig. 6: Block diagram of the energy harvesting system

Photovoltaic devices are a mature commercial product, and offer the availability of high energy density per area. They are however, limited to ambient lighting and temperature conditions. For our system, An IXOLAR XOD17-04B Solar Cell was chosen as the input to the power IC system. This power IC was used as a pulse resonant power converter designed to be self-powered and self-controlled for maximum power point tracking and low switching loss. It converted an input voltage of 0.8-1.2 V to a steady 2 V output voltage which was used to power the hydrogen sensor, microcontroller, and RF transmitter. The solar power system was tested in a lab condition using a flash light as the energy source, and was successful in delivering a sufficient amount of power to power the microcontroller, sensor interface, and RF transmitter.

For the harvesting of vibrational energy, piezoelectric devices are attractive as sources in that no light source is required, and the collection of energy is proportional to the volume of the devices. The limiting factor, however, is the magnitude and frequency of the vibrations. Four PSI D220-A4-203YB double quick mounted Y-pole benders were used as the PZT device, and were connected to a direct charging circuit using a full-bridge rectifier and a shunt capacitor. By adjusting the loading mass on the PZT, the resonant frequency of the PZT bimorph device was adjusted. The vibrational energy harvesting system was tested using a mechanical shaker tuned to 1 grms @ 130 Hz. It successfully delivered 250 µW to the hydrogen sensor node.

A power analysis was performed to examine the power requirements of the system. The power consumption required for the microcontroller to remain idle, output data via serial power, and to scan the A/D converter input, was a constant power of 2.5 µW. The most power consumed by the microcontroller at any time, was in the microcontroller's initialization state, which occured only once during initial power up of the microcontroller. The initialization time for the microcontroller was only 12.5 ms, where average power consumption was 3.07 mW with a peak power of 7.3 mW, as shown in Fig. 7. With the Ming TX-99 transmitter attached to the serial output port of the microcontroller, the RF transmitter module did not consume power for the RF transmission of a logic "0", and an average power of 261 µW with a peak of 522 µW was consumed for the transmission of a logic "1" with 500 us pulse width, as shown in Fig. 8. The charging and discharging characteristics of Figure 8 were due to the LC resonant tank which was used to set the operating frequency of our RF transmitter to 300 MHz.

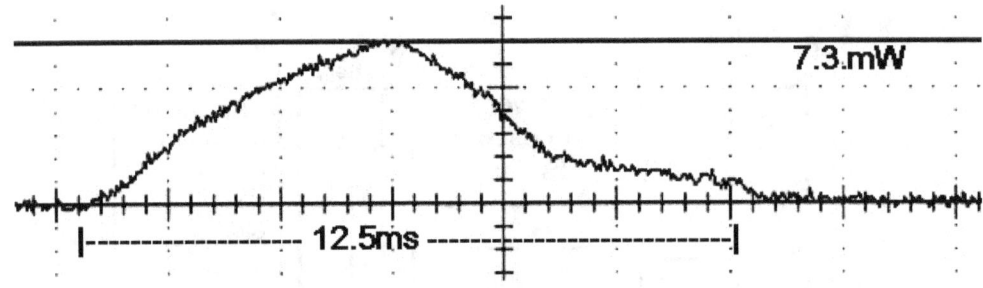

Fig. 7: DC power consumption of microcontroller during initialization

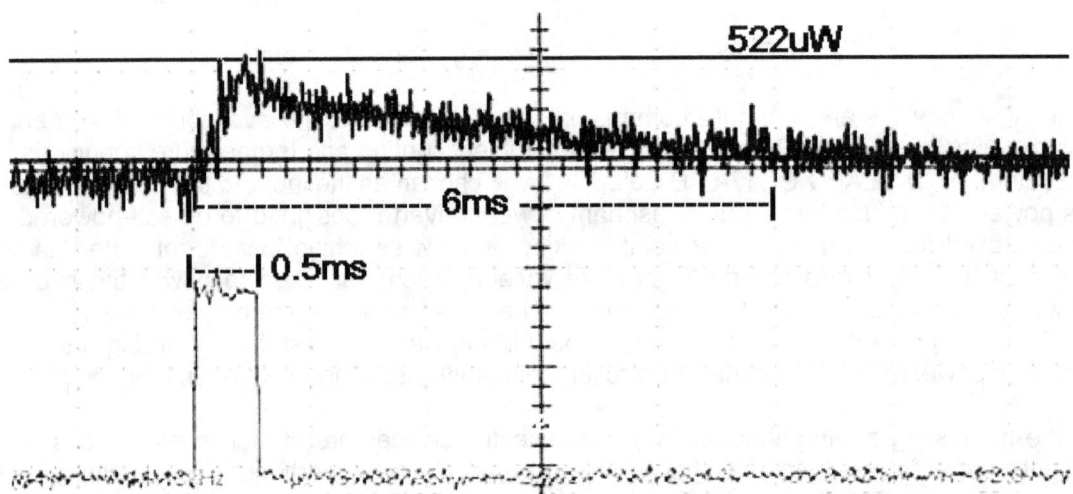

Fig. 8: DC power consumption of RF transmitter when transmitting bit "1" (top trace) activated by input data of 500 μs pulse width (lower trace)

A power analysis was also performed for the sensor interface. This power analysis included both the power to drive the instrumentation amplifier, as well as the power to bias circuit for the resistance bridge. With no hydrogen present, the sensor interface consumed 84 μW of power. With 10 ppm and 500 ppm of hydrogen exposure, the interface consumes 88.4 μW and 88.6 μW, respectively, as listed in Table 1.

Field Test of Low Power Wireless Hydrogen Sensor Network
Fully integrated low-power sensor modules with the sensor device, detection circuit, control unit, and wireless transceiver have been fabricated. An IEEE 802.15.4 WPAN (Zigbee) compliant 2.4 GHz wireless sensor network has been set up for data transmission, to accommodate a number of hydrogen sensor nodes implemented in the system. In addition, a user friendly program installed on the wireless network server has been developed to send real time data of each individual hydrogen sensor via the internet, so that the data can be analyzed and monitored from anywhere that has Internet connection. An alarm function alerting respective personnel of a potential hydrogen leak by sending the message through the cellular phone network has also been successfully implemented. Field tests have also been conducted to show the effectiveness and stability of the whole system.

To achieve the goal of detecting reactions due to hydrogen only and excluding other changes caused by variables such as temperature and moisture, a differential detection interface was used. Several kinds of differential devices have been fabricated and each of its performance has been evaluated to select the most effective solution. These differential devices have two sensors integrated on the same chip. The two sensors are identical except one is designed to react to hydrogen whereas the other one is covered by dielectric protection layer and not exposed to ambient gas. Fig. 9(a) shows the die photo of a differential sensor device, and Fig. 9(b) shows the test result using 1% hydrogen. One sensor reacted promptly while the other had no significant response, which functions as expected.

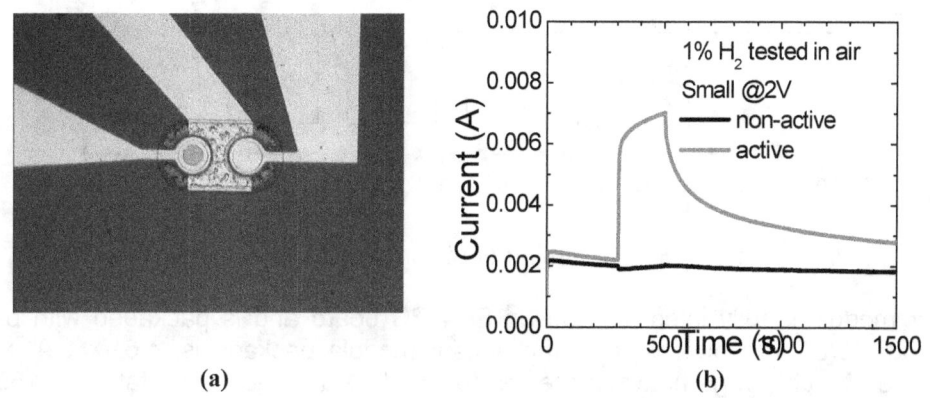

Fig. 9: Die photo (a) and response to 1% H_2 (b) of differential GaAs sensor device

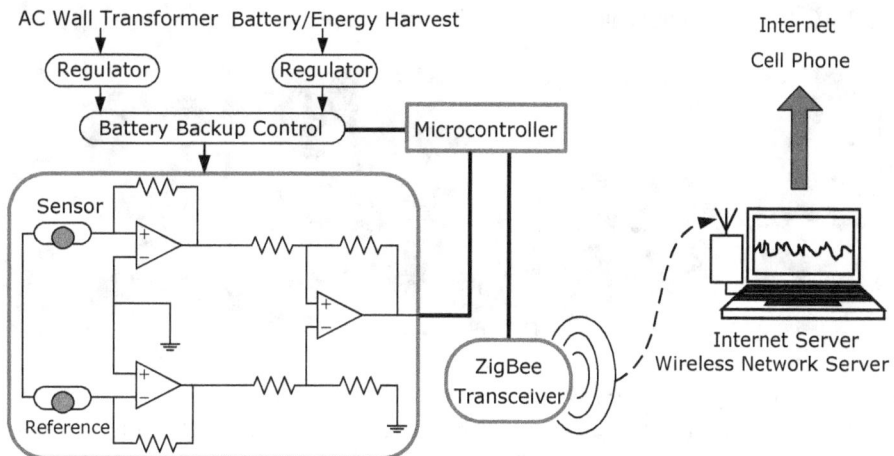

Fig. 10: Block diagram of sensor module and wireless network server

An instrumentation amplifier is used by the detection circuit to sense the change of current in the device. The current variation, embodied as a change in the output voltage of the detection circuit, is fed into the microcontroller. The microcontroller calculates the corresponding current change and controls the transceiver to transmit the data to the wireless network server. The block diagram of the sensor module and the wireless network server are shown in Fig. 10.

The Zigbee compliant wireless network supports the unique needs of low-cost, low-power sensor networks, and operates within the ISM 2.4 GHz frequency band. The transceiver module is completely turned down for most of the time, and is turned on to transmit data in extremely short intervals. The timing of the system is shown in Fig. 11. When the sensor module is turned on, it is programmed to power up for the first 30 sec. Following the initialization process, the detection circuit is periodically powered down for 5 sec and powered up again for another 1 sec, achieving a 16.67% duty cycle. The ZigBee transceiver is enabled for 5.5 ms to transmit the data only at the end of every cycle. This gives a RF duty cycle of only 0.09% and significantly saves the power consumption.

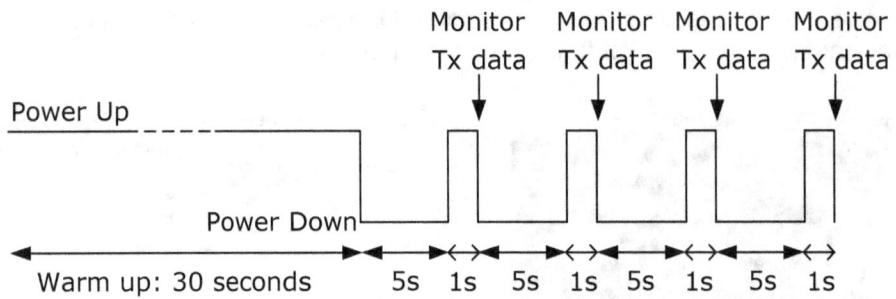

Fig. 11: System timing of at wireless sensor node.

The sensor module is fully integrated on an FR4 PC board and is packaged with battery as shown in Fig. 12(a). The dimension of the sensor module package is: 4.5" × 2.9" × 2". The maximum line of sight range between the sensor module and the base station is 150 m. The base station of the wireless sensor network server is also integrated in a single module (3.0" × 2.7" × 1.1") and is ready to be connected to a laptop by a USB cable, as shown in Fig. 12(b) and (c). The base station draws its power from the laptop's USB interface, thus do not require any battery or wall AC transformer, which reduces its form factor.

Fig. 12: Photos of the sensor system:
(a) sensor with sensor device, (b) sensor and base station, (c) computer interface with base station.

Field tests have been conducted at both University of Florida in Gainesville, Florida, and Greenway Ford in Orlando, Florida. Fig. 13 (a) and (b) show the pictures of installing wireless hydrogen sensors on the ceilings in these two facilities, respectively.

(a)   **(b)**

Fig. 13: Installation of sensors for field tests in (a) University of Florida (b) and Greenway Ford

Several sensor modules were tested in University of Florida before installation in Greenway Ford. Fig. 14 shows the test result of six sensors. Each sensor was exposed to 1% hydrogen gas for three times and was monitored continuously for four hours after the test.

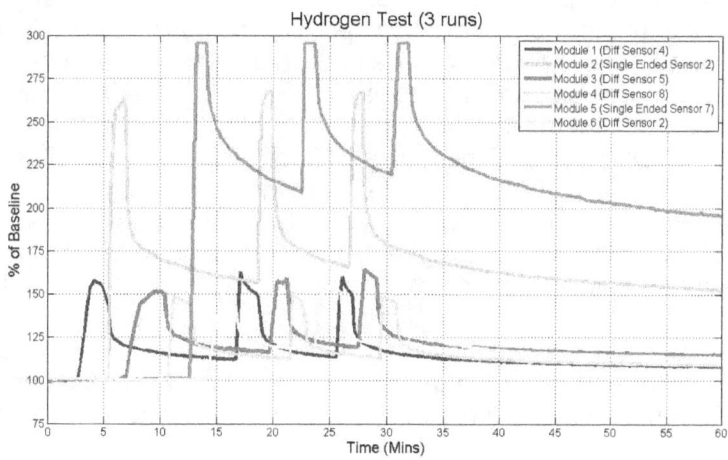

Fig. 14: Test result of six hydrogen sensor modules

A hydrogen sensor server software was developed using .NET Framework v3.5. It has the functionalities of communication port setting, emergency alarm, data collection, and real-time data plot viewing. The interfaces of "General" and "Detail" sections are show in Fig. 15 and Fig. 16.

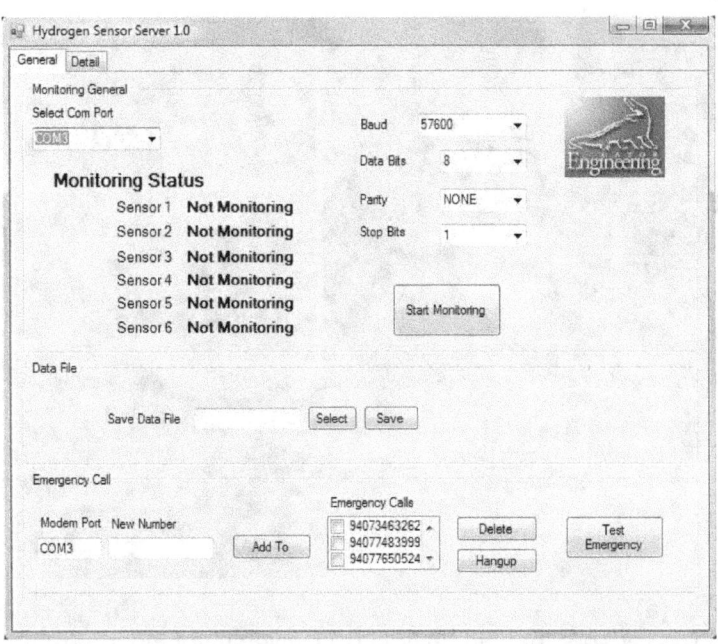

Fig. 15: Hydrogen sensor server General section

Fig. 15 shows the "General" section. It is separated into three subareas: Monitoring General, Data File, and Emergency Call. "Monitoring General" subarea handles the clients' need to configure the communication port and monitor the sensor status in brief. Clients select the communication port and related communication specifications based on their computer systems which are sometimes different. A group of default parameters have been preset for clients' convenience. Clients can turn on or off the monitoring by clicking the "Start Monitoring" button. A brief status report is presented under "Monitoring Status". The status of the sensor is either "Not monitoring", "Normal", "Offline", or "Emergency". "Data File" subarea takes care of the data collection function. Clients simply select the data file he needs and save it to the destination disk. More sophisticated remote data collection using ASP.NET technology will be presented later in this report. "Emergency Call" subarea brings the clients an efficient and adjustable emergency reporting mechanism. In case of emergency, the server will make phone calls to the numbers listed in the "Emergency Calls" list (A modem and phone line connected to the server computer is required). Clients can add or delete the numbers in the phone list. Also, since the emergency situation is rare, we provide a "Test Emergency" function for clients to test the emergency calls. The emergency situation is also monitored by the remote network server at University of Florida.

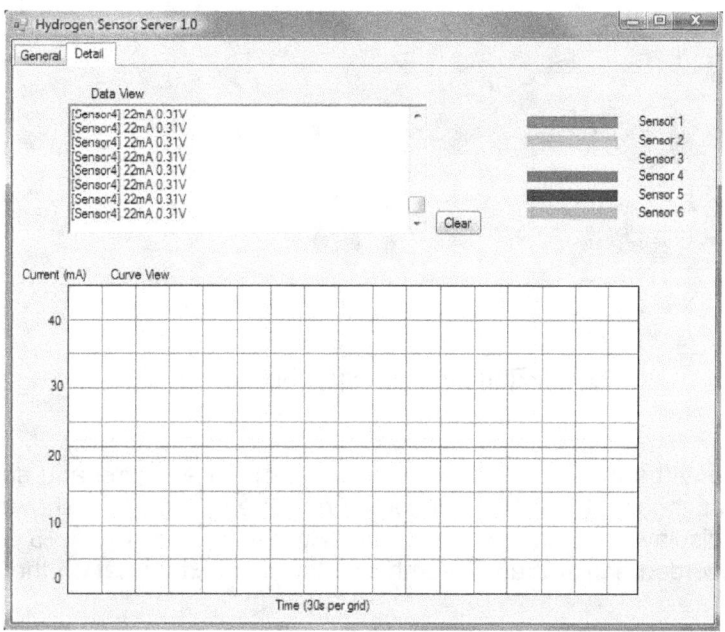

Fig. 16: Hydrogen sensor server Detail section

Fig. 16 shows the "Detail" section. It is separated into two subareas: Data View and Curve View. Only one sensor in constant hydrogen density is turned on for this test view, though the software is capable of displaying the data of all sensors. In "Data View" subarea, the data channeled through the ZigBee receiver is reported by three parameters: Sensor ID, sensed current, and sensed voltage. These three parameters are also transmitted through Internet to the network server at University of Florida for remote monitoring purposes. The "Curve View" presents the data in a graphical plot. The curves of six sensors use different colors. This curve view will also be presented in the remote sensing website for real-time monitoring.

A remote hydrogen sensing system was developed to present the data plot to Internet users, regardless of the user locations. Fig. 17 shows the remote system design. Hydrogen sensor server near the sensor module location sends the data in individual packages to the web server at University of Florida. The web server saves the data package on daily basis in text format. The system makes use of the data package format to avoid the data loss that sometimes occurs on the Internet. IIS (Internet Information Service) coming with the web server channels the data package to the ASP.NET program. ASP.NET program analyzes the data and generates a web page.

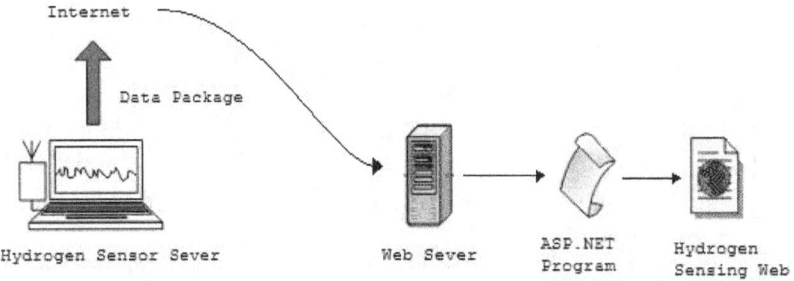

Fig. 17: Remote Hydrogen Sensing System

Internet users around the world can access the web page at any time and see the plotted data exactly the same as those on the local sensor server. The web page is shown in Fig. 18. Users can select the display time range including Real-time, 85 min, 15 hr, 6 days, and 2 months. Data recorded earlier than 2 months is stored in text format on the web server and is not plotted.

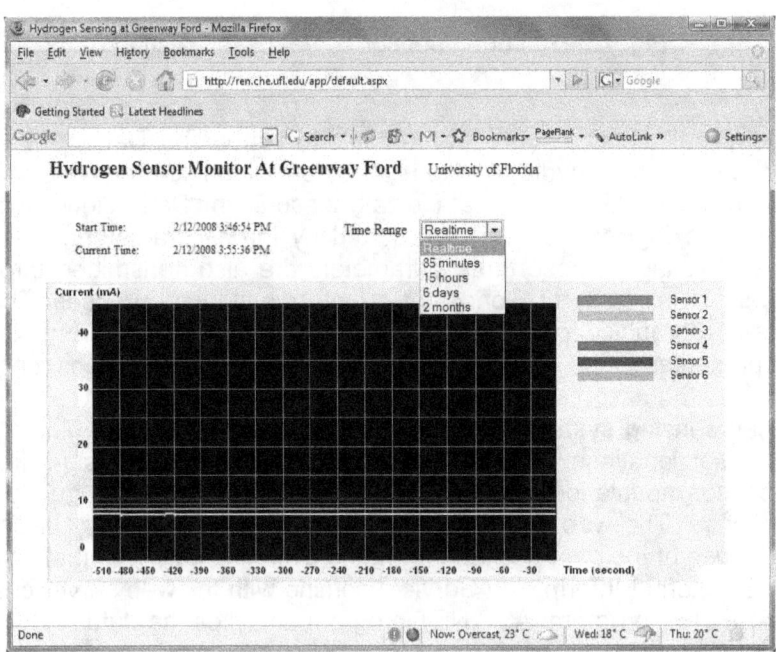

Fig. 18: Remote Hydrogen Sensing Webpage

One hydrogen sensor base station was set up at the Greenway Ford. Fig. 19 shows the base station setup. Six sensor modules were installed near the hydrogen vehicle service bay.

Fig. 19: Hydrogen Sensor Server Setup

There are four people in the emergency call list, including two general managers of Greenway Ford, the chief IT technician of Greenway Ford, and the programmer of the hydrogen sensor server. The emergency call function was tested successfully on the day of installation. Field test at Greenway Ford has been conducted successfully since installation.

Conclusions

A hydrogen detecting system consisting of solar and vibrational energy harvesting devices, ZnO nano-rod based sensor, sensor interface, microcontroller, and the RF front end of our sensor system was designed and fabricated. The system was capable of detecting the presence of hydrogen and transmitting the detected sensor data, while obtaining the necessary power for operation through energy scavenging devices.

A wireless sensor network which meets the IEEE 802.15.4 standards has been constructed to transmit data from a number of hydrogen sensors to a base station. A hydrogen sensor server software was developed using the latest .NET Framework v3.5 technology. The sensor server can be installed and launched easily to perform sensing tasks without the need of previously used MATLAB program. The user interface was refined to be more user-friendly when performing functions such as communication port setting, emergency alarm, data collection, and data plot. The entire system has been tested for functionality and stability at the University of Florida and Greenway Ford in Orlando. Results from field tests indicate that the low-power hydrogen sensor can work stably and react quickly to detect hydrogen leak. In addition, a hydrogen sensing ASP.NET website was set up for users to view the hydrogen sensor data in different time windows including the most important real-time display option. The collected data will be very useful for future analysis and demonstration of hydrogen safety.

Patents

1. USPTO Patent Cooperation Treaty Application: System for Hydrogen Sensing, UF Invention Disclosure #12267 (Aug. 3, 2006), filed on Oct. 5, 2006.

Publications
Journals
1. T. Anderson, H. T. Wang, C. Li, Z. N. Low, B. S. Kang, J. Lin, S. J. Pearton, A. Osinsky, Amir Dabiran, P. Chow, J. Painter, F. Ren, "A New Advance in Hydrogen Sensors," (Invited) *Hydrogen and Fuel Cell Safety*, July 2007.
2. J. Jun, B. Chou, J. Lin, A. Phipps, S. Xu, K. Ngo, D. Johnson, A. Kasyap, T. Nishida, H. T. Wang, B. S. Kang, T. Anderson, F. Ren, L. C. Tien, P. W. Sadik, D. P. Norton, L. F. Voss, and S. J. Pearton, "A Hydrogen Leakage Detection System Using Self-Powered Wireless Hydrogen Sensor Nodes," *Solid State Electronics*, Vol. 51, Issue 7, pp. 1018-1022, July 2007.
3. H. T. Wang, T. J. Anderson, B. S. Kang, F. Ren, C. Li, Z. N. Low, J. Lin, B. P. Gila, S. J. Pearton, A. Osinsky, A. Dabiran, " Stable hydrogen sensors from AlGaN/GaN heterostructure diodes with TiB_2-based Ohmic contacts," Applied Physics Letters, 90, 252109, June 2007.
4. Y. Yoon, J. Lin, S. J. Pearton, J. Guo, "Role of Grain Boundaries in ZnO Nanowire Field-Effect Transistors," *Journal of Applied Physics*, Vol. 101, Issue 2, 024301 (5 pages), January 15, 2007.
5. H. T. Wang, T. J. Anderson, F. Ren, C. Li, Z. N. Low, J. Lin, B. P. Gila, S. J. Pearton, A. Osinsky, A. Dabiran, "Robust Detection of Hydrogen Using Differential AlGaN/GaN High Electron Mobility Transistor Sensing Diodes," *Applied Physics Letters*, 89, 242111, December 2006.
6. B. S. Kang, H. T. Wang, L. C. Tien, F. Ren, B. P. Gila, D. P. Norton, C. R. Abernathy, J. Lin, and S. J. Pearton, "Wide Bandgap Semiconductor Nanorod and Thin Film Gas Sensors," *Sensors*, Vol. 6, No. 6, pp. 643-666, June 2006.
7. L. C. Tien, P. W. Sadik, D. P. Norton, L. F. Voss, S. J. Pearton, H. T. Wang, B. S. Kang, F. Ren, J. Jun, and J. Lin, "Hydrogen sensing at room temperature with Pt-coated ZnO thin films and nanorods," *Applied Physics Letters*, 87, 222106, November 2005.
8. S. N. G. Chu, F. Ren, S. J. Pearton, B. S. Kang, S. Kim, B. P. Gila, C. R. Abernathy, J.-I. Chyi, W. J. Johnson and J. Lin, "Piezoelectric polarization-induced two-dimensional electron gases in AlGaN/GaN heteroepitaxial structures: Application for micro-pressure sensors," *Materials Science and Engineering: A*, Vol. 409/1-2, pp. 340-347, November 2005.
9. H. T. Wang, B. S. Kang, F. Ren, L. C. Tien, P. W. Sadik, D. P. Norton, S. J. Pearton, and J. Lin, "Detection of hydrogen at room temperature with catalyst-coated multiple ZnO nanorods," *Applied Physics A*, Vol. 81, No. 6, pp. 1117-1119, November 2005.
10. L. C. Tien, H. T. Wang, B. S. Kang, F. Ren, P. W. Sadik, D. P. Norton, S. J. Pearton, and J. Lin, "Room-Temperature Hydrogen-Selective Sensing Using Single Pt-Coated ZnO Nanowires at Microwatt Power Levels," *Electrochemical and Solid-State Letters*, 8 (9), G230-G232, July 2005.
11. H. T. Wang, B. S. Kang, F. Ren, L. C. Tien, P. W. Sadik, D. P. Norton, S. J. Pearton, and J. Lin, "Hydrogen-selective sensing at room temperature with ZnO nanorods," *Applied Physics Letters*, 86, 243503, June 2005.
12. A. EL. Kouche, J. Lin, M. E. Law, S. Kim, B. S. Kim, F. Ren, S. J. Pearton, "Remote Sensing System for Hydrogen Using GaN Schottky Diodes," *Journal of Sensors and Actuators B: Chemical*, Vol. 105/2, pp. 329-333, 2005.
13. S. J. Pearton, B. S. Kang, Suku Kim, F. Ren, B. P. Gila, C. R. Abernathy, J. Lin, and S. N. G. Chu, "GaN-based diodes and transistors for chemical, gas, biological and pressure sensing," *J. Phys.: Condensed Matter*, Vol. 16, Issue 29, pp. R961–R994, July 2004.

Conference Papers

1. S. Pearton, F. Ren, B. Kang, H. Wang, B. Gila, D. Norton, L. Tien, T. Chancellor, T. Lele, Y. Tseng, J. Lin, "GaN and ZnO-Based Sensors for Gas, Nuclear Materials and Chemical Detection," accepted, *Proceedings of the E10 Symposium "Wide-Bandgap Semiconductor Materials & Devices 8," of the 212th Meeting of the Electrochemical Society*, Washington, D.C., October 7-12, 2007.
2. L. Tien, D. Norton, B. Kang, H. Wang, F. Ren, J. Lin, S. Pearton, "ZnO Nanowires for Sensing and Device Applications," accepted, *Proceedings of the E6 Symposium "Nanoscale One-Dimensional Electronic and Photonic Devices," of the 212th Meeting of the Electrochemical Society*, Washington, D.C., October 7-12, 2007.
3. H. Wang, T. Anderson, F. Ren, C. Li, Z. Low, J. Lin, B. Gila, S. Pearton, A. Dabiran and A. Osinsky, "Robust Detection of Hydrogen Using Differential AlGaN/GaN High Electron Mobility Transistor Sensing Diodes," *Proceedings of the STATE-OF-THE-ART PROGRAMS ON COMPOUND SEMICONDUCTOR at the 211th Meeting of the Electrochemical Society*, May 2007.
4. S. J. Pearton, L. C. Tien, H. S. Kim, D. P. Norton, J. J. Chen, H. T. Wang, B. S. Kang, F. Ren, W. T. Lim, J. Wright, R. Khanna, L. F. Voss, L. Stafford, J. Jun, J. Lin, "Development of Thin Film and Nanorod ZnO-Based LEDs and Sensors," in Materials Research Society Symposium Proceedings Vol. 957, K-01-05, 12 pages, December 2006.
5. T. Nishida, J. Lin, K. Ngo, F. Ren, D. Norton, S. Pearton, L. Cattafesta, M. Sheplak, J. Jun, A. Kasyap, D. Johnson, and A. Phipps, "Wireless Hydrogen Sensor Self-powered Using Ambient Vibration and Light," *Proceedings of 2006 ASME International Mechanical Engineering Congress and Exposition* (IMECE), 6 pages, November 2006.
6. K. Ngo, T. Nishida, J. Lin, S. Xu, and A. Phipps, "Power Converters for Piezoelectric Energy Extraction," *Proceedings of 2006 ASME International Mechanical Engineering Congress and Exposition* (IMECE), 7 pages, November 2006.
7. J. Jun, J. Lin, S. Xu, A. Phipps, K. Ngo, D. Johnson, A. Kasyap, T. Nishida, H. T. Wang, B. S. Kang, F. Ren, L. C. Tien, P. W. Sadik, D. P. Norton, L. F. Voss and S. J. Pearton, " Low-Power Detection of Hydrogen Leakage Using a Self-Powered Wireless Hydrogen Sensor Node," *Proceedings of the AIChE 2006 Spring National Meeting*, 10 pages, April 2006.
8. H. T. Wang, B. S. Kang, F. Ren, J. Jun, J. Lin, L. C. Tien, P. W. Sadik, D. P. Norton, L. F. Voss, S. J. Pearton, "Highly sensitive hydrogen sensor using Pt nanoparticles coated ZnO single and multiple nanowires," *Proceedings of 208th Meeting of the Electrochemical Society*, pp. 238-247, 2005.
9. H. T. Wang, B. S. Kang, F. Ren, R. C. Fitch, J. K. Gillespie, N. Moser, G. Jessen, T. Jenkins, R. Dettmer, D. Via, A. Crespo, J. Lin, B. P. Gila, C. R. Abernathy, L. C. Tien, D. P. Norton, S. J. Pearton, "Hydrogen-induced reversible changes in drain current of Pt-gated AlGaN/GaN HEMTs," *Proceedings of 208th Meeting of the Electrochemical Society*, pp. 274-283, 2005.
10. J. Lin, A. EL Kouche, M. E. Law, F. Ren, B. S. Kang, S. J. Pearton, D. P. Norton, and C. R. Abernathy, "GaN-Based and Zno Nanorod Sensors for Wireless Hydrogen Leak Detection," *Proceedings of the STATE-OF-THE-ART PROGRAMS ON COMPOUND SEMICONDUCTOR at the 207th Meeting of the Electrochemical Society*, 12 pages, 2005. (Invited)
11. A. EL Kouche, J. Lin, M. E. Law, S. Kim, B. S. Kim, F. Ren, S. J. Pearton, "Remote Sensing System for Hydrogen Detection Using GaN Schottky Diodes," *Proceedings of the 6th IEEE Wireless and Microwave Technology Conference*, p. 76, 2004.

Presentations

1. Y. Wang, W. Lim, L. Covert, T. Anderson, J. Lin, S. Pearton, D. Norton and F. Ren, "Room Temperature Deposited Enhancement Mode and Depletion Mode Indium Zinc Oxide Thin Film Transistors," *E7 Symposium "ZnO, InZnO, and InGaO Related Materials and Devices for Electronic and Photonic Applications," of the 213th Meeting of the Electrochemical Society*, Phoenix, AZ, May 20, 2008. (Refereed conference abstract)
2. Y. Wang, F. Ren, L. Covert, J. Lin, W. Lim, S. Pearton, "Frequency Response and Devices Performance of the Indium Zinc Oxide Thin Film Transistors," *E8 Symposium "State-of-the-Art Program on Compound Semiconductors (SOTAPOCS 47)," of the 212th Meeting of the Electrochemical Society*, Washington, D.C., October 7-12, 2007. (Refereed conference abstract)
3. T. Anderson, H. T. Wang, B. S. Kang, C. Li, Z. N. Low, J. Lin, S. J. Pearton, J. Painter, C. Balaban, A. Osinsky, A. Dabiran, P. Chow, and F. Ren, "Advances in Wireless Hydrogen Sensor Networks," NHA Annual Hydrogen Conference 2008, March 31-April 4, 2008. (Refereed conference abstract)
4. T. Anderson, H. T. Wang, B. S. Kang, F. Ren, C. Li, Z. N. Low, J. Lin, and S. J. Pearton, "Wireless Hydrogen Sensor Networks Using GaN-based Devices," NHA Annual Hydrogen Conference 2007, San Antonio, TX, March 19-22, 2007. (Refereed conference abstract)
5. J. Lin, "Self-Powered Wireless Nano-Sensor for Hydrogen Leak Detection and Wireless Power Transmission," Radio Science Symposium for A Sustainable Humanosphere, Kyoto, Japan, March 20-21, 2006. (Invited)
6. J. Lin, A. EL. Kouche, M. E. Law, F. Ren, B. S. Kang, S. J. Pearton, D. P. Norton, and C. R. Abernathy, "Challenges of Building a Wireless Hydrogen Sensor Network," Florida Chapter of the AVS Science and Technology Society Annual Joint Symposium, Orlando, Florida, March 13-17, 2005. (Invited)
7. J. Lin, "Wireless Sensor System" in Energy Colloquium, University of Florida, November 1, 2004. (Invited)

Students from Research

1. Ahmad (Ed) El Kouche (BS in EE 2003, MS in CISE 2006)
2. Jerry Jun (MS in EE May 2006) – Motorola
3. Bruce Chou (MS in EE May 2006) – Intel
4. Kwok Wai Mak (BS in EE December 2006)
5. Changzhi Li (PhD student)
6. Zhen Ning Low (PhD student)
7. Xiaogang Yu (PhD student)

Funding Obtained by Leveraging NASA Grant

1. Hydrogen Sensing System, $14,000, Florida Department of Environmental Protection, 8/15/2006-3/31/2007
2. Hydrogen Sensor System Prototype for Ford Motor Company, $25,000 University of Florida Office of Technology Licensing, 8/15/2006-3/31/2007

Collaborations

Dr. T. Nishida, Dr. K. Ngo (now with Virginia Tech), Dr. J. Guo – Department of Electrical and Computer Engineering
Dr. F. Ren – Department of Chemical Engineering
Dr. S. Pearton, Dr. D. Norton – Department of Material Science and Engineering
Mr. J. Painter – Ford Greenway in Orlando

Acknowledgements

Technical discussions and suggestions from Dr. Gary Hunter of NASA Glenn Research Center are greatly appreciated. The support from Florida State Department of Environmental Protection, Greenway Ford, and Mr. John Painter are crucial to the field test of the system.

References

[1] G. W. Hunter, P. G. Neudeck, R. S. Okojie, G. M. Beheim, V. Thomas, L. Chen, D. Lukco, C. C. Liu, B. Ward and D. Makel, Proc. ECS Vol. 01-02 212(2002).
[2] L. Y. Chen, G. W. Hunter, P. G. Neudeck, D. L. Knight, C. C. Liu, and Q. H. Wu, Proceeding of the Third International Symposium on Ceramic Sensors, H. U. Anderson, M. Liu, and N. Yamazoe, Editors, Electrochemical Society Inc. Pennington, NJ, pp. 92-98, (1996).
[3] G. W. Hunter, C.C. Liu, D. Makel, MEMS Handbook, ed. M. G. Hak (CRC Press, Boca Raton, 2001).
[4] L. Chen, G. W. Hunter, and P. G. Neudeck, J. Vac. Sci. Technol A 15, 1228 (1997); J. Vac. Sci. Technol. A16 2890(1998).
[5] Q. Wan, Q. H. Li, Y.J. Chen, T. H. Wang, X. L. He, X. G. Gao and J. P. Li , Appl. Phys. Lett. 84 3654(2004).
[6] K. Keem, H. Kim, G. T. Kim, J. S. Lee, B. Min, K. Cho, M. Y. Sung and S. Kim, Appl. Phys. Lett. 84 4376(2004).
[7] M. H. Huang, S., Mao, H. Feick, H. Yan, Y. Wu, H. Kind, E. Weber, R. Russo and P. Yang, Science 292 1897 (2001).
[8] Z. L. Wang, Materials Today, pp. 26-33,June 2004.
[9] Y. W. Heo, D. P. Norton, L. C. Tien, Y. Kwon, B. S. Kang, F. Ren, S. J. Pearton and J. R. LaRoche, Mat. Sci. Eng. R 47,1(2004).
[10] C. H. Liu, W. C. Liu, F. C. K. Au, J. X. Ding, C. S. Lee and S. T. Lee, Appl. Phys. Lett. 83, 3168(2003).
[11] W. I. Park, G. C. Yi, J. W. Kim and S. M. Park, Appl. Phys. Lett.82 4358(2003).

2. Novel ZnO Nanorod Hydrogen Gas Sensors

Task PI: Dr. Fan Ren, Chemical Engineering, University of Florida

Collaborators: Dr. Steve Pearton, Materials Science & Engineering, Dr. David P. Norton, Materials Science & Engineering and Dr. Jenshan Lin, Electrical and Computer Engineering, University of Florida

Research Period: August 3, 2004 to March 31, 2008

Abstract

We have demonstrated room temperature hydrogen sensors with wide energy bandgap ZnO semiconductor nanowire devices. The nanowires are contacted at each end by Ohmic contacts and coated with discontinuous Pt films to act as catalysts for decomposing the hydrogen molecules. We have demonstrated selective hydrogen detection with both multiple nanowires and also even lower power operation using single nanowires. The trade-off in the latter case is the increased number of fabrication steps.

This report includes Dr. David Norton's work in the area of Material Growth and Material Characterization. Dr. Fan Ren's research in this collaborative effort is in Sensor Fabrication and Testing.

Introduction

There is strong current interest in the development of lightweight hydrogen sensors capable of ppm sensitivity and extended operation at low power levels. The applications for these sensors include combustion gas detection in spacecraft and solid oxide fuel cells with proton-exchange membranes (PEM). Structures such as nanorods and nanotubes are natural candidates for these applications. To enhance the detection sensitivity for hydrogen, the use of catalytic Pt or Pd coatings or doping on semiconductors (either films or nanorods) and carbon nanotubes can increase the dissociation efficiency of molecular hydrogen to the more reactive atomic form. ZnO is attractive for sensing applications because of its wide bandgap (3.2 eV), the availability of heterostructures, the ease of synthesizing nanostructures and the bio-safe characteristics of this material. A number of previous reports have shown that ZnO with Pd nanoparticles can increase the detection sensitivity for H_2. The high surface-volume ratio of nanorods makes them attractive for detecting hydrogen at low concentrations and it is relatively straightforward to synthesize ZnO nanostructures on a wide variety of substrates. However, prior to this project, there had been no clear demonstration of improved detection sensitivity with nanorods compared to thin films.

Results and Discussion

ZnO nanorods grown by site selective Molecular Beam Epitaxy show current-voltage characteristics that are sensitive to the presence of hydrogen or ozone in the measurement ambient at room temperature for H_2. The sensitivity to hydrogen increases sharply with temperature and multiple nanorods contacted at both ends by Ohmic electrodes show a differential current change of ~18% when changing from a pure N_2 ambient to 10% H_2 in N_2. The nanowire sensors show very low power operation and have a simple fabrication approach.

(i) Hydrogen Sensing Using Multiple ZnO Nanorods

The site selective growth of the nanorods was achieved by Molecular Beam Epitaxy. The growth time was ~2 h at 600 °C. The typical length of the resultant nanorods was 2~10 µm, with typical diameters in the range of 30–150 nm. Selected area diffraction patterns showed the nanorods to be single-crystal. Figure 1 shows a transmission electron micrograph of a single ZnO nanorod. E-beam lithography was used to pattern sputtered Al/Ti/Au electrodes contacting both ends of multiple nanorods on the Al_2O_3 substrates using a shadow mask. The separation of the electrodes was ~3 um. A scanning electron micrograph of the completed device is shown in Figure 2. Au wires were bonded to the contact pad for current –voltage (I-V) measurements performed over the range 25-150 °C in 10%H_2 in N_2. Note that no currents were measured through the discontinuous Au islands and no thin film of ZnO was observed with the growth condition of nanorods. The I-V characteristics from the multiple nanorods measured in either N_2 or 10H_2/90% N_2 ambients at different temperatures. At room temperature there was no detectable change in current but the presence of the hydrogen in the ambient can be detected beginning at ~112 °C. The reversible chemisorption of reactive gases at the surface of these metal oxides can produce a large and reversible variation in the conductance of the material. The gas sensing mechanism suggested include the desorption of adsorbed surface oxygen and grain boundaries in poly-ZnO, exchange of charges between adsorbed gas species and the ZnO surface leading to changes in depletion depth and changes in surface or grain boundary conduction by gas adsorption/desorption. The detection mechanism is still not firmly established in these devices and needs further study. When detecting hydrogen with the same types of rectifiers, we have observed changes in current consistent with changes in the near -surface doping of the ZnO, in which hydrogen introduces a shallow donor state. The nanorods also showed a strong photoresponse to above bandgap UV light (366 nm). The photoresponse was fast and indicates that the changes in conductivity due to injection of carriers is bulk-related and not due to surface effects.

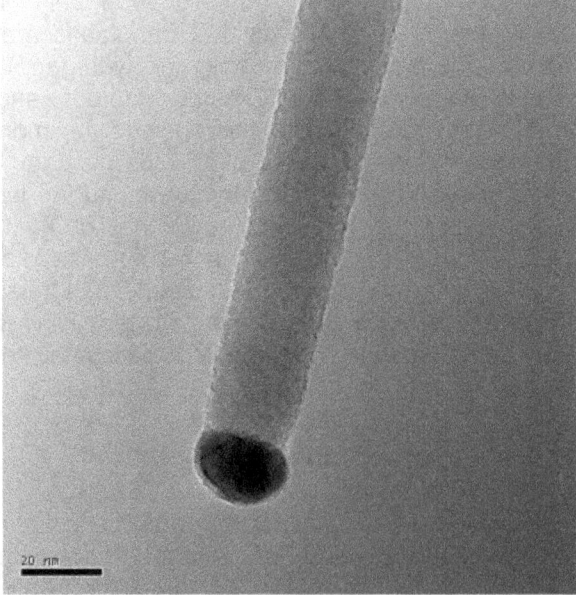

Figure 1. TEM of ZnO nanorod.

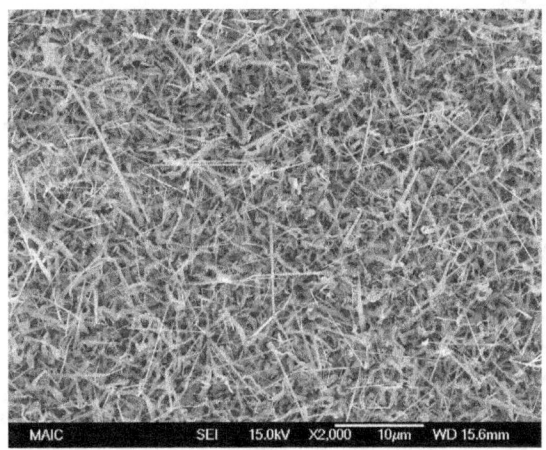 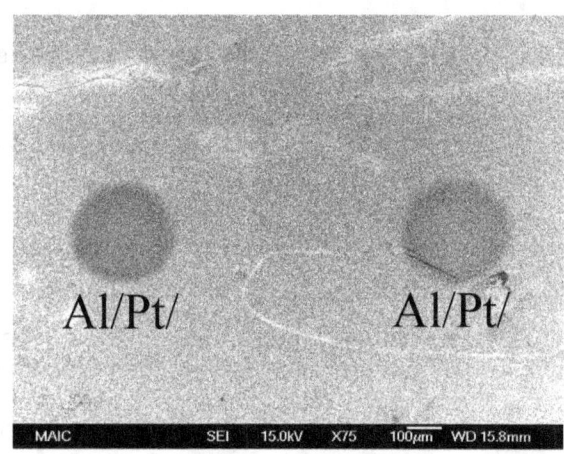

Figure 2. SEM of left) ZnO multiple nanorods and right) the pattern contacted by Al/Pt/Au electrodes.

Figure 3 shows the difference in current at fixed bias of 0.1 V for measurement in N_2 versus 10%H_2/90% N_2 as a function of the measurement temperature. The change in current is still in the nA range at 112 °C but increases with temperature and is about 16 nA at 200 °C. These changes are readily detected by conventional ammeters. However, the inability to detect hydrogen at room temperature means that an on-chip heater would still be needed for any practical application of ZnO nanorods for detection of combustion gases.

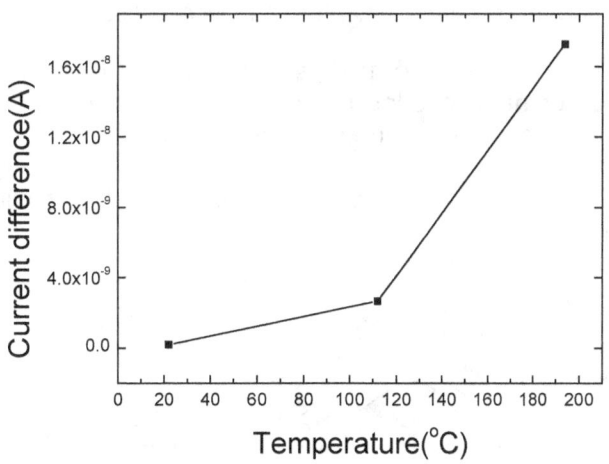

Figure 3. Change in current measured at 0.1 V for measurement in either N_2 or 10H_2 in N_2 ambients.

(ii) Use of metal catalyst layers to increase detection efficiency

A variety of different metal catalyst coatings (Pt, Pd, Au, Ag, Ti, and Ni) deposited on multiple ZnO nanorods were compared for their effectiveness in enhancing sensitivity for detecting hydrogen at room temperature. Pt-coated nanorods show a relative response of up to 8% in room temperature resistance upon exposure to hydrogen concentrations in N_2 of 500 ppm. This is a factor of two larger than obtained with Pd and more than an order of magnitude larger than achieved with the remaining metals. The power levels for these sensors were low, ~0.4 mW for the responses noted above. Pt-coated ZnO nanorods easily detected hydrogen down to 100 ppm, with relative responses of 4% at this concentration after 10 min exposure. The nanorods show a return to their initial conductance upon switching back to a pure air ambient with time constants of order a few minutes at room temperature. This slow response at room temperature is a drawback in some applications but the sensors do offer low power operation and ppm detection sensitivity.

There is currently great interest in the development of hydrogen sensors for applications involving leak detection in hydrogen fuel storage systems and fuel cells for space craft. One of the main demands for such sensors is the ability to selectively detect hydrogen at room temperature in the presence of air. In addition, for most of these applications, the sensors should have very low power requirements and minimal weight. Nanostructures are natural candidates for this type of sensing. We made a comparison of different metal coating layers on multiple ZnO nanorods for enhancing the sensitivity to detection of hydrogen at room temperature. Pt is found to be the most effective catalyst, followed by Pd. The resulting sensors are shown to be capable of detecting hydrogen in the range of ppm at room temperature using very small current and voltage requirements and recover quickly after the source of hydrogen is removed.

The nanorods were coated with Pd, Pt, Au, Ni, Ag or Ti thin films (~100Å thick) deposited by sputtering. The Pd and Pt are known to be the most effective catalysts for dissociation of molecular hydrogen, the Au was chosen to see if it could provide any enhancement in hydrogen sensitivity since it might potentially be used as an over-layer to prevent oxidation of the other metals which are all significantly cheaper than Pd and Pt and were explored from the viewpoint of keeping the overall cost of the sensor fabrication as low as possible.

Contacts to the multiple nanorods were formed using a shadow mask and sputtering of Al/Ti/Au electrodes. The separation of the electrodes was ~30 um. A schematic of the resulting sensor is shown in Figure 4. Au wires were bonded to the contact pad for current –voltage (I-V) measurements performed at 25 °C in a range of different ambients (N_2, O_2 or 10-500 ppm H_2 in N_2). The I-V characteristics from the multiple nanorods were linear with typical currents of 0.8 mA at an applied bias of 0.5 V.

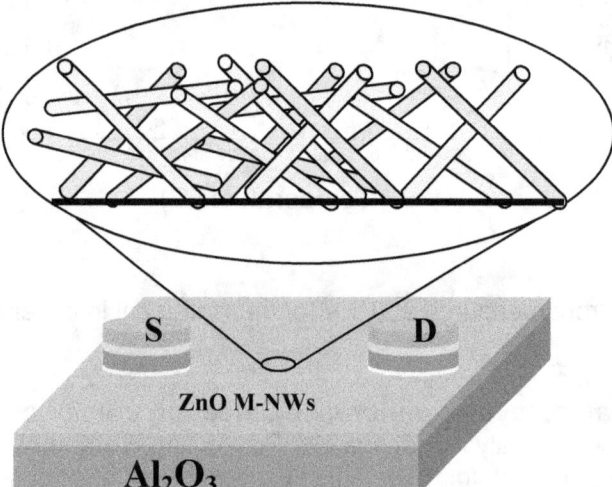

Figure 4. Schematic of contact geometry for multiple nanorod gas sensor.

Figure 5 shows the time dependence of relative resistance change of either metal-coated or uncoated multiple ZnO nanorods as the gas ambient is switched from N_2 to 500 ppm of H_2 in air and then back to N_2 as time proceeds. These were measured a bias voltage of 0.5V. The first point of note is that there is a strong increase (approximately a factor of 5) in the response of the Pt-coated nanorods to hydrogen relative to the uncoated devices. The maximum response was ~8%. There is also a strong enhancement in response with Pd coatings, but the other metals produce little or no change. This is consistent with the known catalytic properties of these metals for hydrogen dissociation. Pd has a higher permeability than Pt but the solubility of H_2 is larger in the former. Moreover, studies of the bonding of H to Ni, Pd and Pt surfaces have shown that the adsorption energy is lowest on Pt. There was no response of either type of nanorod to the presence of O_2 in the ambient at room temperature. Once the hydrogen is removed from the ambient, the recovery of the initial resistance is rapid (<20 sec). By sharp contrast, upon introduction of the hydrogen, the effective nanorod resistance continues to change for periods of >15 mins. This suggests that the kinetics of the chemisorption of molecular hydrogen onto the metal and its dissociation to atomic hydrogen are the rate-limiting steps in the resulting change in conductance of the ZnO. An activation energy of 12 KJ/mole was calculated from a plot of the rate of change of nanorod resistance. This energy is somewhat larger than that of a typical diffusion process and suggests the rate-limiting step mechanism for this sensing process is more likely to be the chemisorption of hydrogen on the Pd surface. Finally, Figure 5 shows an incubation time for response of the sensors to hydrogen. This could be due to some of the Pd becoming covered with native oxide which is removed by exposure to hydrogen. A potential solution is use a bi-layer deposition of the Pd followed by a very thin Au layer to protect the Pd from oxidation. However this adds to the complexity and cost of the process and since the Pd is a not a continuous film, the optimum coverage of Au would need to be determined. We should also point out that the I-V characteristics were the same when measured in vacuum as in air, indicating that the sensors are not sensitive to humidity.

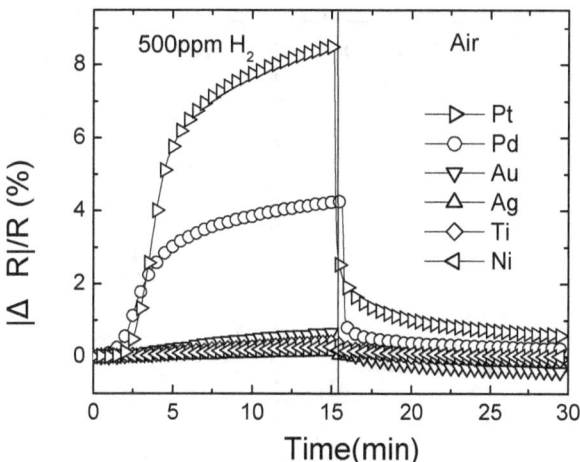

Figure 5. Time dependence of relative resistance response of metal-coated multiple ZnO nanorods as the gas ambient is switched from N_2 to 500 ppm of H_2 in air as time proceeds. There was no response to O_2.

The power requirements for the sensors were very low. Figure 6 shows the I-V characteristics measured at 25 °C in both a pure N_2 ambient and after 15 mins in a 500 ppm H_2 in N_2 ambient. Under these conditions, the resistance response is 8% and is achieved for a power requirement of only 0.4 mW. This compares well with competing nanotechnologies for hydrogen detection such as Pd-loaded carbon nanotubes. Moreover, the 8% response compares very well to the existing SiC-based sensors, which operate at temperatures >100 °C through an on-chip heater in order to enhance the hydrogen dissociation efficiency.

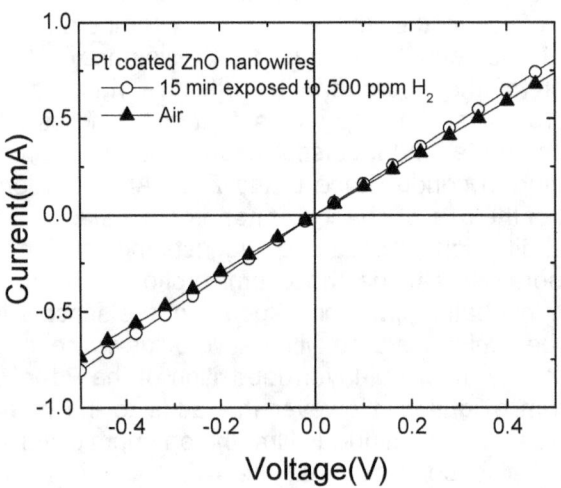

Figure 6. I-V characteristic of Pt-coated nanowires in air and after 15 mins in 500 ppm H_2 in air.

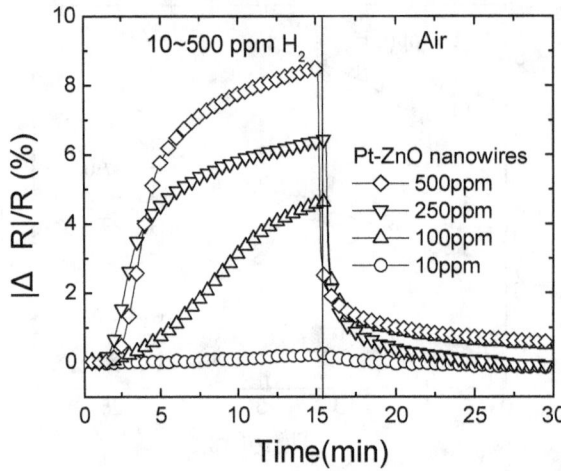

Figure 7. Time dependence of resistance change of Pd-coated multiple ZnO nanorods as the gas ambient is switched from N_2 to various concentrations of H_2 in air (10-500 ppm) and then back to N_2.

Figure 7 shows the sensors can detect 100 ppm H_2 in N_2 at room temperature. In conclusion, Pt-coated ZnO nanorods appear well-suited to detection of ppm concentrations of hydrogen at room temperature. The recovery characteristics are fast upon removal of hydrogen from the ambient. The ZnO nanorods can be placed on cheap transparent substrates such as glass, making them attractive for low-cost sensing applications and can operate at very low power conditions. Of course, there are many issues still to be addressed, in particular on the reliability and long-term reproducibility of the sensor response before it can be considered for space-flight applications. In addition, the slow response of the sensors at room temperature is a major issue in some applications.

(iii) Comparison of the sensitivities for detecting hydrogen with Pt-coated single ZnO nanorods and thin films.

The Pt-coated single nanorods show a current response approximately a factor of three larger at room temperature upon exposure to 500 ppm H_2 in N_2 than the thin films of ZnO. The power consumption with both types of sensors can be very small (in the nW range) when using discontinuous coatings of Pt. Once the Pt coating becomes continuous, the current required to operate the sensors increases to the µW range. The optimum ZnO thin film thickness under our conditions was between 40-170 nm, with the hydrogen sensitivity falling off outside this range. The nanorod sensors show a slower recovery in air after hydrogen exposure than the thin films, but exhibit a faster response to hydrogen, consistent with the notion that the former adsorb relatively more hydrogen on their surface. Both ZnO thin and nanorods can not detect oxygen.

Two types of ZnO were employed in these experiments. The thin films were grown by Pulsed Laser Deposition on sapphire substrates at 450 °C. The ZnO thickness was varied from 20-350 nm. The films were nominally undoped with low n-type (10^{17} cm^{-3}) carrier concentration. Top-side Ohmic contacts of sputtered Al/Ti/Au were patterned by lift-off. The site-selective growth of ZnO nanorods was achieved by nucleating the nanorods on a substrate coated with Au islands as has also been described in detail previously. For nominal Au thicknesses of 20Å, discontinuous Au islands are realized. ZnO nanorods were deposited by Molecular Beam Epitaxy (MBE). The typical length of the resultant nanorods was 5~15 µm, with typical diameters in the range of 50~150 nm. Selected area diffraction patterns showed the nanorods to be single crystal. For the multiple nanorods, a shadow mask was used to pattern sputtered Al/Ti/Au electrodes on the ZnO nanorods/Al_2O_3 substrates. The separation of the electrodes was ~400 µm. In some cases, the sensors were coated with Pt thin films (~10Å thick) deposited by sputtering. Au wires were bonded to the contact pad for current–voltage (I-V) measurements performed at 25 °C in air, N_2 or 500 ppm H_2 in N_2. No currents were measured through the discontinuous Au islands.

Figure 8 shows the I-V characteristics measured between the Ohmic contacts on the thin film ZnO samples of either 20 or 350 nm thickness, both before and after the Pt deposition on the surface. The current increase as a result of the Pt deposition is approximately a factor of two for the thinnest sample and remains in the nA range at 0.5 V bias, i.e., the power consumption is 4 nW at this operating voltage. The effective conductivity of the Pt-coated films is higher due to the presence of the metal. At longer Pt sputtering times we would typically see a transition to much higher currents as the Pt film became continuous and the conductivity of the structure was no longer determined by the ZnO layer itself.

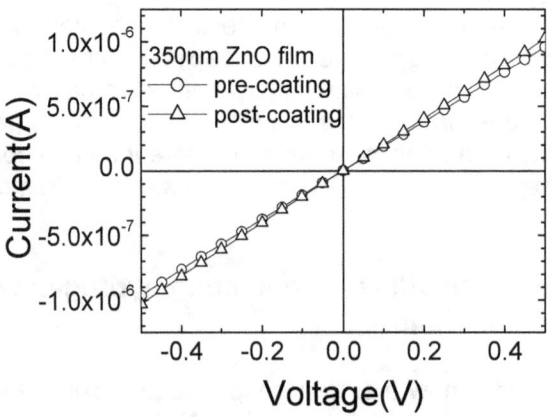

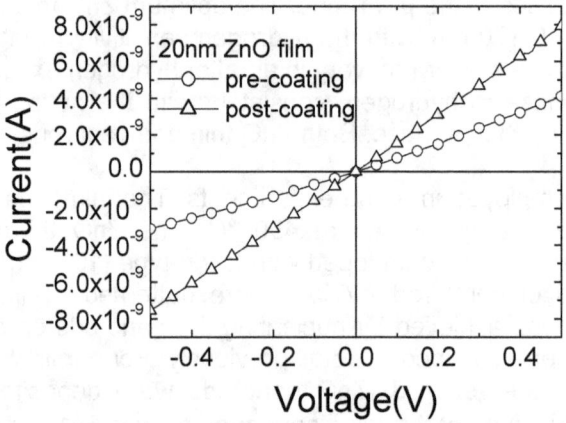

Figure 8. Room temperature I-V characteristics from ZnO thin films of thickness 20 or 350 nm measured in air before and after coating with Pt.

Figure 9 (top) shows the time dependence of current change at 0.5 V bias on the Pt-coated ZnO films of different thickness as the gas ambient is switched from N_2 to 500 ppm H_2 in N_2 and back to air as time proceeds. This data shows that the sensors are insensitive to N_2 and that there is a strong ZnO thickness dependence to the response to hydrogen. The bottom of Figure 9 shows the change in current at 0.5 V bias when switching from N_2 to the hydrogen-containing ambient for the ZnO films of different thickness. At small thicknesses, the current change is small, which is probably related to poorer crystal quality and also at large film thickness where the bulk conductivity dominates the total resistance.

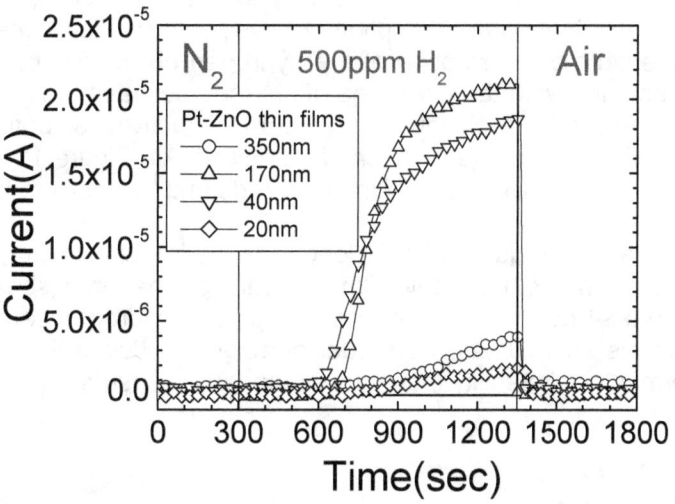

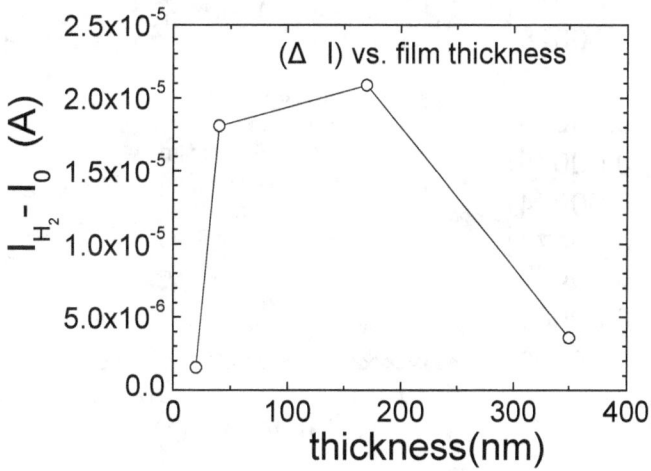

Figure 9. Current as a function of time for Pt-coated ZnO thin films of different thickness cycled from N_2 to 500 ppm H_2 in N_2 to air ambient (top) and change in current at fixed bias (0.5 V) when switching to the H_2-containing ambient (bottom).

Figure 10 shows the time dependence of current in both the Pt-coated multiple ZnO nanorods and the thin films as the gas ambient is switched from N_2 to 500 ppm H_2 in N_2 and then back to air as time proceeds. It is clear that the nanorods have a much larger response (roughly a factor of 3 even for the optimal response for the thin films) to the introduction of hydrogen into the ambient compared to their thin film counterparts. This is consistent with the expectation of a higher relative response based on their larger surface-to-volume ratio. Although not shown here, there was no response of either type of sensors to the presence of O_2 in the ambient at room temperature. The recovery of the initial resistance is rapid (90%, <20 sec) upon removal of the hydrogen from the ambient by either O_2 or air, while the nanorod resistance is still changing at least 15 mins after the introduction of the hydrogen. The response is faster at higher temperatures. The nanorods show a slower recovery than the thin films, most likely due to the relatively higher degree of hydrogen adsorption. The expected sensing mechanism suggested previously is that reversible chemisorption of the hydrogen on the ZnO produces a reversible variation in the conductance, with the exchange of charges between the hydrogen and the ZnO surface leading to changes in depletion depth. The conductivity of both ZnO thin film and nanorods did change when the ambient switched from N_2 to Air. Figure 11 shows the maximum current change at 0.5 V bias for exposure of the nanorods and thin films to the 500 ppm H_2 in N_2.

As discussed earlier, a key requirement in long-term hydrogen sensing applications is the sensor power consumption. Both the thin film and multiple nanorod sensors can operate at ≤0.5 V bias and powers ≤4 nW. We have also demonstrated hydrogen sensing with single ZnO nanorods at power levels approximately an order of magnitude lower than this, but the devices show poorer long-term current stability than multiple nanorod sensors.

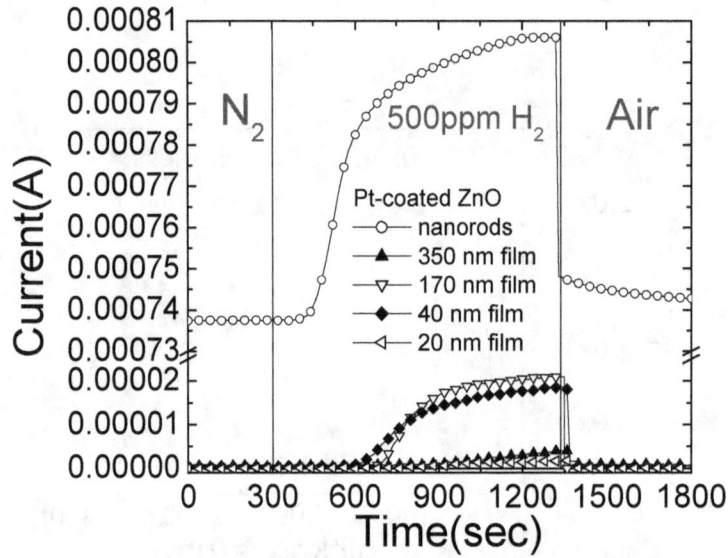

Figure 10. Time dependence of current from Pt-coated ZnO nanorods and thin films nanorods as the gas ambient is switched from N_2 to 500 ppm H_2 in N_2, then to air for recovery.

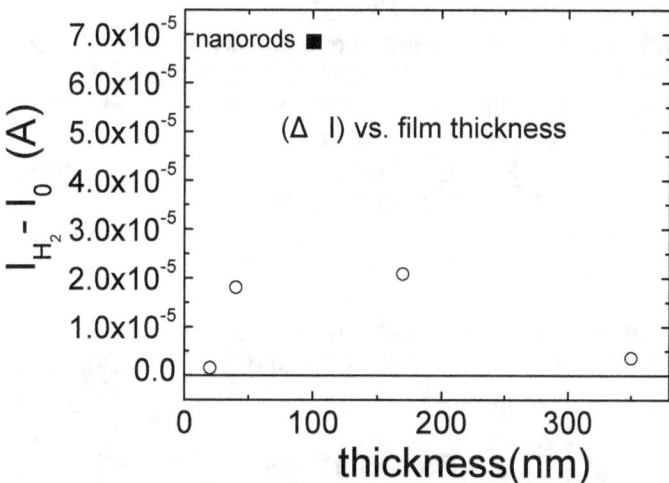

Figure 11. Change in current at fixed bias (0.5 V) when switching to the H_2-containing ambient of either Pt-coated ZnO nanorods or thin films as the gas ambient is switched from N_2 to 500 ppm H_2 in N_2, then to air for recovery.

Conclusions

In conclusion, Pt-coated ZnO thin films and multiple nanorods both are capable of detection of ppm concentrations of hydrogen at room temperature. The thin films show optimum responses to the presence of hydrogen at moderate thicknesses. The nanorods show larger responses to hydrogen than the thin films, consistent with their large surface-to-volume ratios and have the advantage in terms of flexibility of the choice of substrate.

Patents

None

Publications

1. "Low temperature (<100 °C) patterned growth of ZnO nanorods arrays on Si", B.S. Kang, S.J. Pearton and F. Ren, Appl.Phys.Lett.90, 083104 (2007).
2. "Detection of hydrogen with SnO_2-coated ZnO nanorods", L.C. Tien, D.P. Norton, B.P. Gila, S.J. Pearton, Hung-Ta Wang, B.S. Kang and F. Ren, Appl. Surf. Sci. 253, 4748 (2007).
3. "Nucleation control for ZnO nanorods grown by catalyst-driven molecular beam epitaxy", L.C. Tien, D.P. Norton, S.J. Pearton, Hung-Ta Wang and F. Ren, Appl. Surf. Sci. 253, 4620 (2007).
4. "Wide Bandgap Semiconductor Nanorod and Thin Film Gas Sensors", B. S. Kang, H.-T. Wang, L.-C. Tien, F. Ren, B. P. Gila, D. P. Norton, C. R. Abernathy, J. Lin and S.J. Pearton, Sensors 6, 643 (2006).

5. "Hydrogen sensing at room temperature with Pt-coated ZnO thin films and nanorods", L. Tien, P. Sadik, D. P. Norton, L. Voss, S. J. Pearton, H. T. Wang, B. S. Kang, F. Ren, J. Jun and J .Lin, Appl. Phys. Lett.87,222106 (2005).
6. "Room temperature hydrogen selective sensing using single Pt-coated ZnO nanowires at microwatt power levels", L. Tien, H. T. Wang, B. S. Kang, F. Ren, P.W. Sadik, D. P. Norton, S. J. Pearton and J. Lin, Electrochem. Solid-State Lett.8 G239(2005).
7. "Hydrogen-selective sensing at room temperature with ZnO nanorods", H. T. Wang, B. S. Kang, F. Ren, L. C. Tien, P. W. Sadik, D. P. Norton, S. J. Pearton, and Jenshan Lin, Appl. Phys. Lett. 86, 243503 (2005).
8. "Fabrication approaches to ZnO nanowire devices", J. R. LaRoche, Y. W. Heo, B. S. Kang, L. Tien, Y. Kwon, D. P. Norton, B. P. Gila, F. Ren and S. J. Pearton, J. Electron. Mater. 34 404 (2005).
9. "Hydrogen and ozone gas sensing using multiple ZnO nanorods," B. S. Kang, Y. W. Heo, L. C. Tien, D. P. Norton, F. Ren, B. P. Gila and S. J. Pearton, Appl. Phys.A 80 1029 (2005).

Presentations

1. "Low Temperature (<100 °C) Patterned Growth of ZnO Nanorod Arrays on Si," by B. Kang, S. Pearton and F. Ren, 211th Meeting of the Electrochemical Society, Chicago, Illinois May 2007.
2. "Wide Bandgap Semiconductor Nanowires for Sensing Applications," by S. Pearton, D. Norton, F. Ren, L. Tien, B. Kang and G. Chi , 211th Meeting of the Electrochemical Society, Chicago, Illinois May 2007.
3. "GaN, ZnO and InN Nanowires for Gas Sensing Systems", S. J. Pearton, D. Norton, F. Ren, B. Gila, B. Kang and L.C. Tien, TMS Annual Meeting and Exhibition, Orlando, FL, Feb 2007.
4. "Chemical Sensing with ZnO Nanorods", D. P. Norton, L. C. Tien, H. T. Wang, P. W. Sadik, B. S. Kang, F. Ren and S. J. Pearton, TMS Annual Meeting and Exhibition, Orlando, FL, Feb 2007.
5. "Development of Thin Film and Nanorod ZnO-Based LEDs and Sensors", S.J. Pearton, W. T Lim, J. S Wright, R. Khanna, L. Voss, L. Stafford, L. Tien, H. S Kim, D. P Norton, J. J Chen, H. T Wang, B. S Kang, F. Ren, J. Jun and Jenshan Lin, MRS Fall Meeting, Boston, Nov 2006.
6. "Highly Selective Hydrogen Sensing with Pt-functionalized ZnO Thin Films and Nanorods", Li-Chia Tien, Hung-Ta Wang, Byoung-Sam Kang, David Norton, Fan Ren and S.J. Pearton, MRS Fall Meeting, Boston, Nov 2006.
7. "Self-Powered Wireless Nano-Sensor for Hydrogen Leak Detection and Wireless Power Transmission", J. Lin, J. Jun, A. Phipps, X. Shengwen, K. Ngo, D. Johnson, A. Kasyap, T. Nishida, H. T. Wang, B. S. Kang, and F. Ren, L. C. Tien, P. W. Sadik, D. P. Norton, L. F. Voss, and S. J. Pearton, Radio Science Symposium for A Sustainable Humanosphere, Kyoto, Japan, March 20-21, 2006.
8. "Wireless Hydrogen Sensor Self-powered Using Ambient Vibration and Light", T. Nishida, J. Lin, K. Ngo, F. Ren, D. Norton, S. Pearton, L. Cattafesta, M. Sheplak, J. Jun, A. Kasyap, D. Johnson, and A. Phipps, 2006 ASME International Mechanical Engineering Congress November 5-10, 2006 - Chicago, Illinois.
9. "Highly Selective Hydrogen Sensing at Room Temperature with Pt-Functionalized ZnO Thin Films and Nanorods", L. C. Tien, P. W. Sadik, D. P. Norton, L. F. Voss and S. J. Pearton, Florida AVS Meeting, Orlando, FL March 2006.

10. "Hydrogen-Selective Sensing at Room Temperature with Pt-Coated ZnO Nanorods", Hung-Ta Wang, Byoung Sam Kang,Fan Ren, Li-Chia Tien, Patrick Sadik,David Norton, S. J. Pearton, Jenshan Lin, 72nd Southeast Section of the APS, Gainesville, FL, Nov 2005.
11. "Highly sensitive hydrogen sensor using Pt nanoparticles coated ZnO single and multiple nanowires", H. T. Wang, B. S. Kang, F. Ren, J. Jun, J. Lin, L. C. Tien, P. W. Sadik, D. P. Norton, L. F. Voss, and S. J. Pearton, 208th ECS Meeting, Los Angeles, CA Oct 2005.
12. "Highly Sensitive Hydrogen Sensor Using Pd Nanoparticles Coated ZnO Nanorods", H.- T. Wang, B. S. Kang, L. Tien, P. Sadik, D. Norton, S.J. Pearton and Fan Ren, 2005 EMC Conference, Santa Barbara, July 2005.

Students from Research

Jon Wright and Travis Anderson were supported from this program. Jon is serving as a summer intern at NASA Glenn in Summer 2007 and will graduate with a Ph.D in Summer 2009. Travis is interning at Sandia National Labs this summer and will graduate with a Ph.D in Spring 2008.

3. Modeling of ZnO Nanorod Hydrogen Gas Sensors

Task PI: Dr. Jing Guo, Electrical & Computer Engineering, University of Florida

Collaborators: Dr. Jenshan Lin, Electrical & Computer Engineering, Dr. Fan Ran, Chemical Engineering, Dr. Steve Pearton, Material Science & Engineering, and Dr. David Norton, Material Science & Engineering, University of Florida

Graduate Students: Youngki Yoon, Electrical & Computer Engineering, University of Florida

Research Period: August 3, 2004 to March 31, 2008

Introduction

The goals of this funding period are to develop a rigorous and sophisticated simulator for ZnO nanowire transistors and sensors. The simulation capabilities for treating essential device physics, including grain boundary effects, self-consistent electrostatics, quantum effects, scattering, need to be established and demonstrated, which are important for simulation of hydrogen sensors based on ZnO nanowires.

Background

Wide bandgap semiconductor nanowires are being extensively explored for nanoscale electronics, optoelectronics, and sensor applications. Field-effect transistors (FETs), light emitters and laser devices, and chemical and biological sensors [5] based on ZnO nanowires have been recently demonstrated.

Experimental

A rigorous device simulator has been developed. Fig. 1 shows the modeled device structure. The intrinsic ZnO channel has variable length, Lch. The source and drain nanowire extensions are n+ doped, and fixed at 200 nm. A coaxial gate is used to control the channel potential. The gate insulator thickness is variable, while the dielectric constant is kept at κ = 10. The flat band voltage is Vfb = -1.5 V. The nanowire diameter is variable. A "square" ring of charge, with variable inner and outer radius, width, and contained charge, is centered with respect to the channel such that it is between the gate and the nanowire. In order to treat electron transport through the nanowire, a mode space approach and an effective mass description (with an electron effective mass of $m^* = 0.38$) are used. A Schrödinger equation is solved at each channel position z for the cross sections perpendicular to the nanowire,

$$\left[-\frac{\hbar^2}{2m^*}\left(\frac{1}{r}\frac{\partial}{\partial r} r \frac{\partial}{\partial r} + \frac{1}{r^2}\frac{\partial^2}{\partial \phi^2} \right) + E_{C,z}(r) \right] \psi_z(r,\phi) = E_{i,z}\psi_z(r,\phi), \qquad (1)$$

where r is the radius position and ϕ is the angle in the polar coordinate system, $\psi_z(r,\phi)$ is the normalized wave function, $E_{C,z}(r)$ is the conduction band edge at the channel position of z as a function of r, and $E_{i,z}$ is the ith subband profile. A drift-diffusion treatment is used along the channel direction for the lowest a few subbands that are relevant to electron transport. The drift diffusion current for the ith subband is computed as,

$$I_i = \mu_n n_i \frac{dE_{i,z}}{dz} + qD_n \frac{dn_i}{dz}, \qquad (2)$$

where ni is the electron density per unit channel length for the ith subband, $D_n = \mu_n(k_B T/q)$ is the diffusion coefficient, and $k_B T/q$ is the thermal voltage. μn = 150 cm^2/Vs is the ZnO crystal mobility, and the value can be smaller for a small diameter nanowires due to surface roughness scattering. However, in these simulations it is assumed to be constant over the range of selected nanowire diameters. The charge ring is assumed to be perfectly parallel to the nanowire with spatially uniformly distributed charge. The conduction band profile in the carrier transport equation must be self-consistently determined with the charge density. A 2D Poisson equation (in r and z) is iteratively solved with the electron transport equation to treat self-consistent electrostatics. The charges due to background dopants, mobile electrons, and the charge ring are considered. The source-drain current is computed once self-consistency between the carrier transport equation and Poisson equation is achieved.

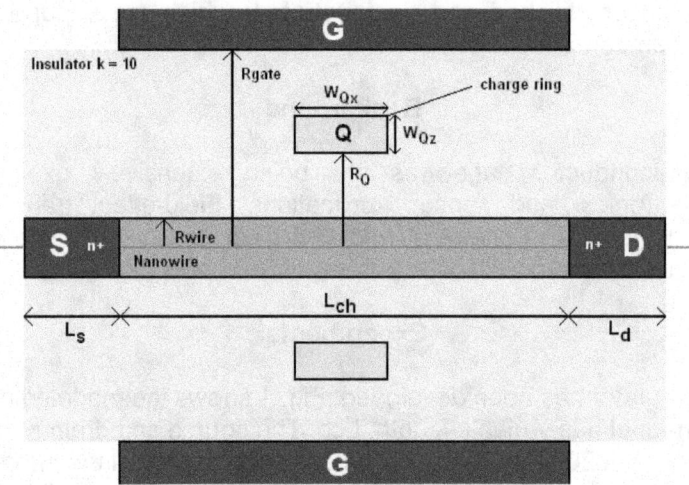

Figure 1. The modeled ZnO nanowire chemical sensor

Results and Discussion

(1) Effect of ZnO nanowire dimensions on sensitivity has been explored using advanced simulation.

The purpose of the simulated sensor is to detect the presence of the ring of charge. To this end, on currents for $V_D = V_G = 1V$ were determined for the case of no charge ring and a single charge ring. The ratio of on-current with to without charge ring present was computed. This ratio, which varies between zero and one, can be considered as an inverse sensitivity. That is, the closer to one the ratio becomes the less sensitive the device is to detecting the presence of the ring of charge. The role of various factors on the on-current sensitivity is examined by solving 2D Schrödinger equation in the nanowire cross section, coupled to a drift-diffusion equation along the nanowire. The presence of a charge ring results in a potential barrier with the thickness determined by the ring thickness and the height determined by the total charge within the ring. As stated, if detected, the charge ring should lead to a decrease of the source-drain current since the voltage drop at the charge ring reduces the electric field at other channel positions.

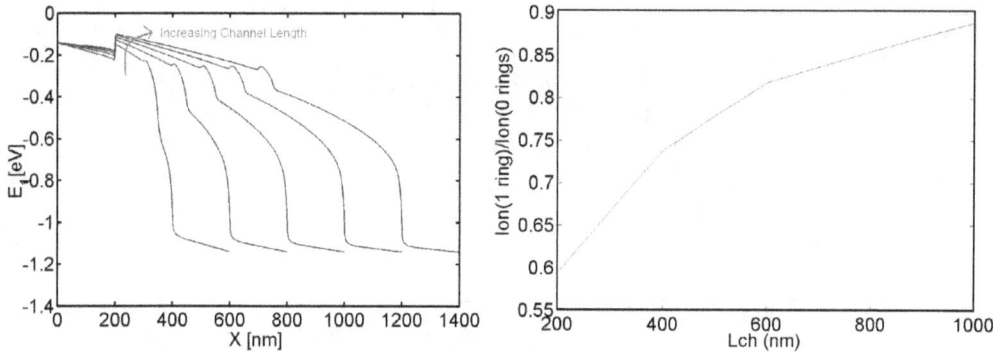

Figure 2. *Channel Length Effect* (a) The lowest subband profile as a function of the channel position. The channel length increases for each curve from left to right, from 200 nm to 1000 nm. A single ring of charge is set in the middle of the channel (b) The ratio of on-currents with to without a ring of charge present. As this value increases, the sensitivity of the sensor decreases.

First, the sensitivity of the device was studied as a function of channel length. The input parameters for this simulation were R_{wire} = 5nm, R_{gate} = 15 nm, W_{Qx} = 50 nm, W_{Qz} = 2.5 nm, Q = 100e-, and R_Q = 7.5 nm (see Figure 1). The channel length, L_{ch}, was set to vary from 200 nm to 1000 nm. Fig. 2a shows how the first conduction subband profile varies as L_{ch} is increased. It is observed that as the length of the channel increases, the effect of the charge becomes less pronounced on the overall bandstructure. Fig. 2b shows the ratio of on-currents as a function of channel length. As previously noted, as this ratio increases, the sensitivity decreases. Thus, it can be inferred that a short channel is highly desirable for this type of sensor.

Next, how wire diameter affects the sensitivity of the device was examined. The input parameters for this simulation were L_{ch} = 1000 nm, R_{gate} = 45 nm, W_{Qx} = 50 nm, W_{Qz} = 2.5 nm, Q = 100e-, and R_Q = 10 nm + R_{wire} (see Figure 1). R_{wire} varies from 2 nm to 10 nm. As mentioned previously, for this simulation it is assumed that the mobility is the same for all considered wire diameters. Fig. 3a (next page) shows the first conduction subband profile, with the wire diameter increasing. It can be seen that as the wire diameter increases, the effect of the charge ring becomes negligible. Fig. 3b shows the ratio of on-currents as a function of wire radius. It is seen that this ratio increases drastically over a small range of radii, and upon reaching approximately .98 when the wire radius is 4 nm, saturates. It should be noted that when the wire radius is 4 nm, the outer radius of the charge ring is 16.5 nm. Since the gate radius is 45 nm, which is almost 3 times the outer radius of the charge ring, it is seen that this is not a result of the gate charge dominating the charge ring.

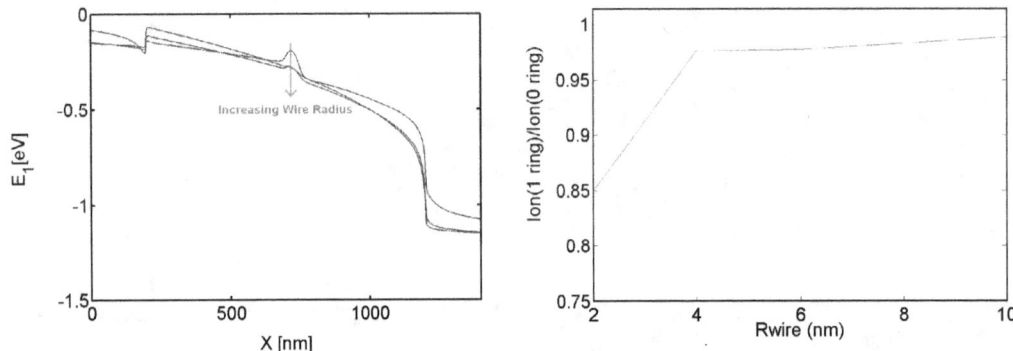

Figure 3. *Wire Radius Effect* (a) The lowest subband profile as a function of the channel position. The wire radius increases for each curve from top to bottom, from 2 nm to 8 nm. A single ring of charge is set in the middle of the channel. (b) The ratio of on-currents with to without a ring of charge present. As this value increases, the sensitivity of the sensor decreases. Note that for sensing a narrow wire is required.

(2) Effect of charge variations on sensitivity has been explored using advanced simulation.

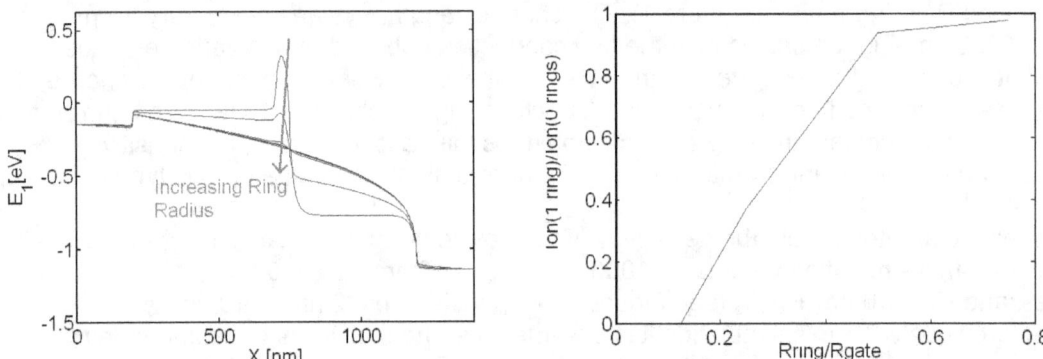

Figure 4. *Charge Ring Radius Effect* (a) The lowest subband profile as a function of the channel position. The charge ring radius increases for each curve from top to bottom, from 5.625 nm to 33.75 nm. The gate radius is 45 nm. A single ring of charge is set in the middle of the channel (b) The ratio of on-currents with to without a ring of charge present. As this value increases, the sensitivity of the sensor decreases. For sensing, a close charge is desired.

A simulation in which the radius of the ring was increased was then performed. For this simulation, the input parameters were L_{ch} = 1000 nm, R_{gate} = 45 nm, W_{Qx} = 50 nm, W_{Qz} = 2.5 nm, Q = 100e-, and R_{wire} = 5 nm. The inner wire radius, R_Q, varies from being equal to .125·R_{gate} to .75·R_{gate}. This method of varying R_Q was selected so that the point at which the gate charge becomes dominant can be observed in terms of the ratio of R_Q to R_{gate}. Fig. 4a shows the effect of increasing the ring radius on the first conduction band. It is clear that as the charge gets further and further from the nanowire surface that the effect on the bandstructure decreases. Fig. 4b shows the ratio of on-currents as a function of charge-ring inner to gate radius. In this figure, it should be noted that the on-current ratio rapidly approaches unity as the ring radius increases until the radius is about half the gate radius. At this point, the increase of the on-current ratio drastically reduces. This is due to the dominance of the gate charge that occurs when the charge ring is closer to the gate than it is to the wire. Thus, this is a gate induced effect.

Next, a simulation was carried out to determine the effect of the width of the charge ring parallel to the channel on the sensor sensitivity. For this simulation, the input parameters were L_{ch} = 1000 nm, R_{gate} = 15 nm, R_Q = 7.5 nm, W_{Qz} = 2.5 nm, Q = 200e-, and R_{wire} = 5 nm. W_{Qx} was varied from 10 nm to 200 nm. Fig. 5a shows the effect that increasing the width has on the first conduction band. It is observed that as the width of the charge ring increases, the effect experienced by the conduction band decreases. Fig. 5b shows the ratio of on-currents as a function of charge ring width. For a 20 nm wide, 200 electron ring, the current is completely eliminated. However, for a 200 nm wide ring with the same total charge, the current only decreases by approximately 15%. This result, coupled with the sensor's heavy dependence on charge proximity, implies that charge orientation is an important factor.

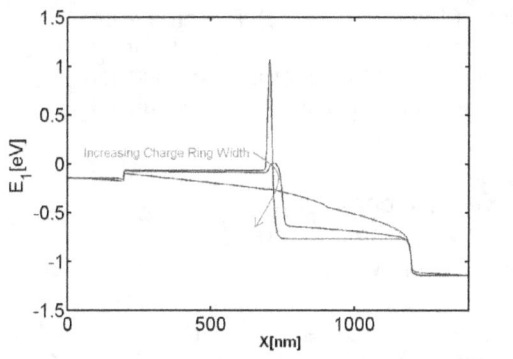 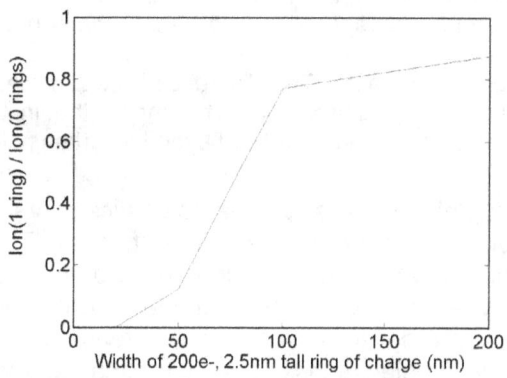

Figure 5. *Charge Ring Width Effect* (a) The lowest subband profile as a function of the channel position. The charge ring width increases for each curve from top to bottom, from 20 nm to 200 nm. The gate radius is 15 nm. A single ring of charge is set in the middle of the channel. (b) The ratio of on-currents with to without a ring of charge present. As this value increases, the sensitivity of the sensor decreases. For sensing, a narrow charge is desired.

A final simulation was carried out to determine the effect that multiple charge rings would have on the sensor. For this simulation, the input parameters were L_{ch} = 1000 nm, R_{gate} = 15 nm, W_{Qx} = 20 nm, W_{Qz} = 2.5 nm, Q = 100e-, R_{wire} = 5 nm, and R_Q = 7.5 nm. The total number of uniformly distributed charge rings was varied from zero to ten. Fig. 6a shows the first conduction band for zero, two, and ten charge rings. From this plot of the E1, it is anticipated that current will drop sharply as a function of the total number of charge rings. This is verified in Fig. 6b, which shows the on-current through the nanowire as a function of the total number of charge rings. It is observed that the current does indeed drop sharply for the first few added charge rings. However, as the total number of charge rings increases, this drop becomes less and less pronounced. This implies that the sensor has some limitations for measuring concentration.

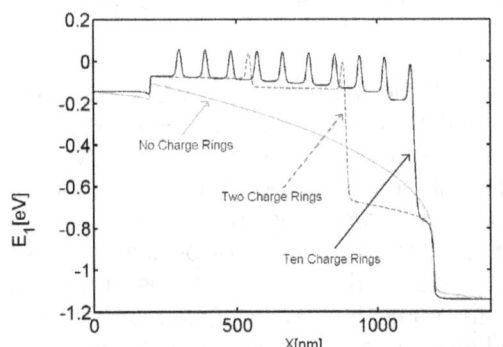 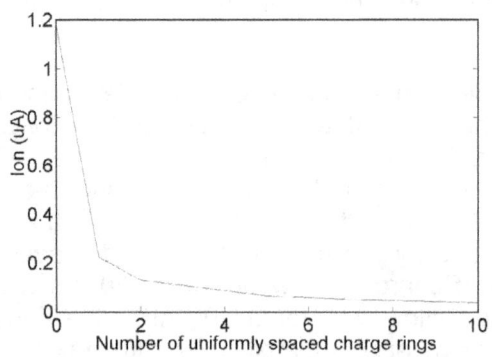

Figure 6. *Number of Charge Rings Effect* (a) The lowest subband profile as a function of the channel position for no, two, and ten rings. The gate radius is 15 nm. Rings of charge are uniformly spaced with respect to the boundaries of the nanowire. (b) The on-current as a function of number of charge rings. As this value decreases, the sensitivity of the sensor increases. The difference between the on current for two charge rings and ten charge rings is very small. This implies that in its present form, this sensor is not suitable to determine concentration beyond a certain point.

Fig. 7a plots the I_D vs. V_G characteristics at V_D = 0.5V for a channel without GBs and a channel with one GB at the middle of the channel. The GB results in a decrease of the source-drain current and an increase of the threshold voltage. Fig. 3b shows the first subband profiles at on-state ($V_D=V_G=0.5V$). A considerable potential drops near the GB, as shown by the solid line in Fig. 7b, in order to overcome the potential barrier at the GB. Because the total potential drops over the whole intrinsic channel remains approximately the same after adding a GB, the potential drop over the GB results in a smaller electric field at other channel positions, as shown in Fig. 3b. The source-drain current, therefore, decreases.

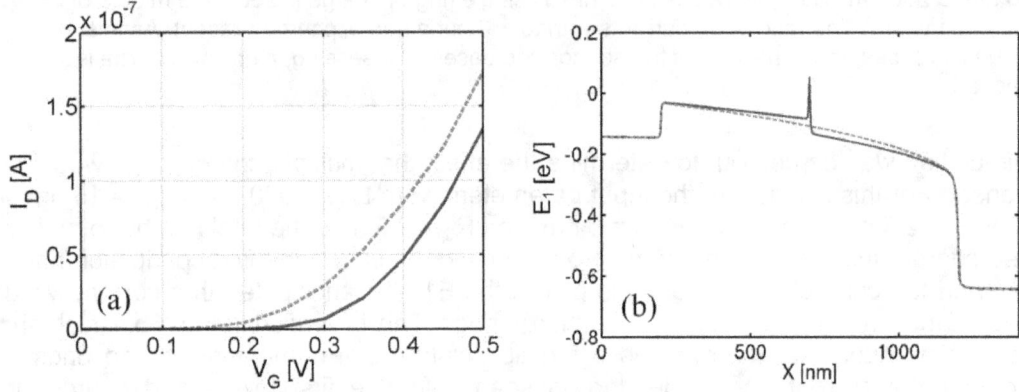

Figure 7. *The effect on I-V characteristics* (a) The I_D vs. V_G at V_D=0.5V and (b) the first subband profile at $V_D=V_G$=0.5V for a single crystal ZnO nanowire channel (the dashed lines) and a channel with one grain boundary (the solid lines). A single grain boundary with a trap density constant of $N_{T0}=8\times10^{13}/cm^2$ exists at the middle of the channel.

Fig. 8a plots the on-current (at $V_G=V_D=0.5V$) as a function of the nanowire diameter. A non-monotonous dependence of the on-current on the wire diameter is observed, which is due to two competing mechanisms. As the nanowires diameter increases, the quantum confinement in the cross section of the nanowires becomes weaker and the higher subband energy decreases, which makes more subbands to deliver current. As the nanowire diameter increases from 5 nm to 10 nm, this effect is dominant and the source-drain current increases. At the same time, the increase of the cross section area of the nanowires results in a larger amount of charge on the GB due to the increase of the GB area. A larger voltage must drop over the GB in order to overcome a higher barrier as shown in Fig. 8b, and thereby the electric field at channel positions away from the GB decreases. For the diameter from 10 nm to 12 nm, this effect is more dominant and the on-current decreases as the diameter increases.

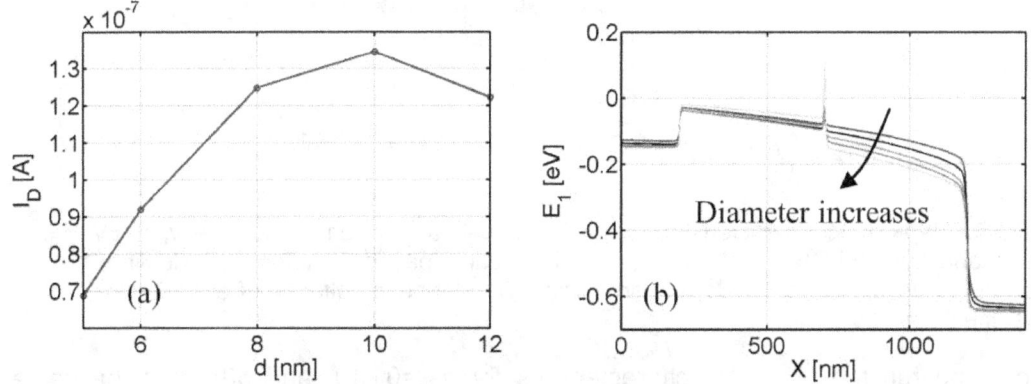

Figure 8. *The effect of the nanowire diameter* (a) The on current (at $V_D=V_G=0.5V$) as a function of the nanowire diameter. (b) The first subband profile for different nanowire diameters d_{wire}=4, 6, 8, 10 and 12 nm at the on-state. A single grain boundary exists at the middle of the channel.

Fig. 9a plots the I_D vs. V_G characteristics at V_D=0.5 for three nanowire channels with different numbers of GBs, n_{GB}=0, 10, and 50 GBs. A trap state density of $N_{T0} = 8 \times 10^{13}$/cm^2 is used. For simplicity, all grain boundaries have the same trap density states and are equally spaced in the intrinsic channel. 10 GBs decreases the on-current by about 72%, and 50 GBs decreases the on-current by about 93%. Increasing the number of GBs results in a larger decrease of the on-current because a larger number of GBs requires a larger potential drop over all GBs. It results in a smaller electric field in the rest part of the channel and therefore a smaller on-current.

Fig. 9b shows the I_D vs. V_G characteristics on a log scale. As the number of the GBs increases, the off-current also decreases significantly and the threshold voltage of the transistor, V_T, increases. (We define V_T as the gate voltage resulting in a source-drain current of 10^{-10}A). Because the applied voltage is typically much smaller than the middle of bandgap of ZnO, no charging and discharging of the trap states occur. Therefore, the subthreshold swing remains unaffected by the existence of the trap states in the simulated gate voltage range.

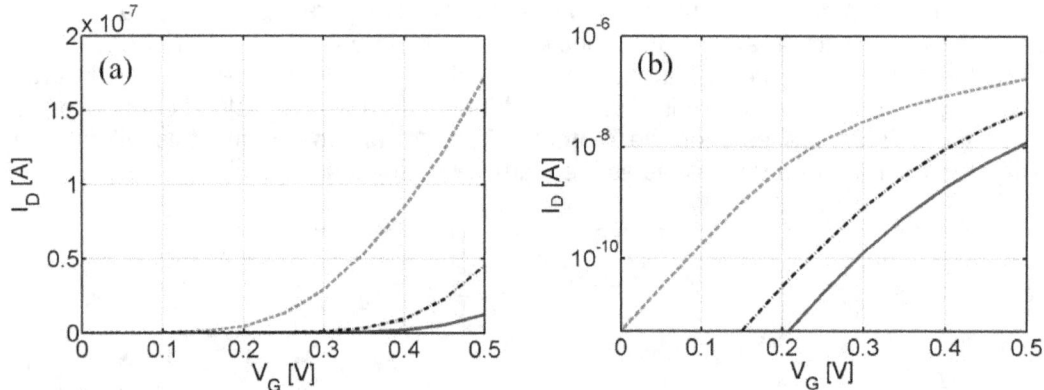

Figure 9. The I_D vs. V_G characteristics (a) on linear scale and (b) on log scale at $V_D=0.5V$ for a channel without GBs (dashed line), with 10 equally spaced GBs (the dash-dot line), and with 50 equally spaced GBs (the solid line). The flat-band voltages of all FETs are the same, $V_{fb}=-1.5V$.

We also compared the I_D vs. V_G characteristics for $n_{GB}=0$, 10, and 50 using the same V_T specification, as shown in Fig. 10. The on-current of the FET with $n_{GB}=10$ is approximately the same as that with $n_{GB}=0$ at the same gate overdrive voltage. Increasing the number of GBs to a larger value of $n_{GB}=50$, however, results in decrease of the source-drain current at the same gate overdrive voltage. If one extracts an effective channel mobility by fitting the I_D vs. V_G characteristics above the threshold voltage using the square law [12], the value extracted for the FET with 50 GBs is about 40% smaller than that without GBs.

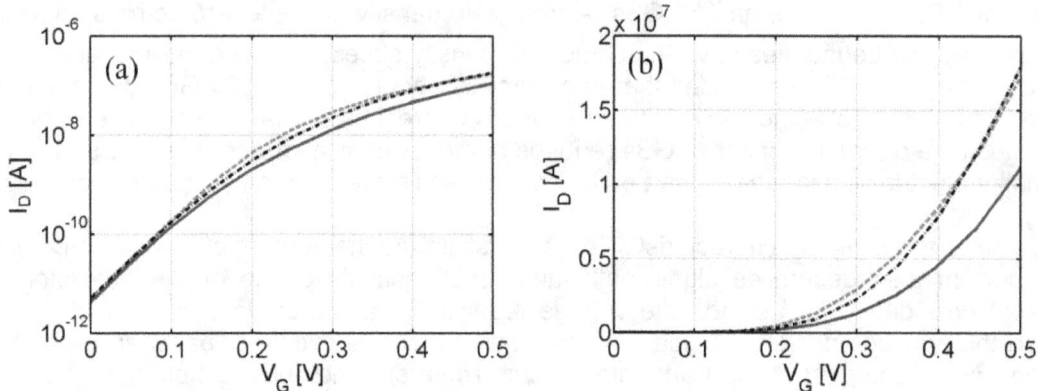

Figure 10. The I_D vs. V_G characteristics with the same V_T specification (a) on the linear scale and (b) on the log scale at $V_D=0.5V$ for a channel without GBs (dashed line), with 10 equally spaced GBs (the dash-dot line), and with 50 equally spaced GBs (the solid line). The threshold voltages of all FETs are adjusted to the same value.

Conclusions

In summary, role of grain boundaries in ZnO nanowire FETs is examined. When the source-drain voltage is applied, the potential drops over the barrier at the GBs result in smaller electric field at other channel positions. Therefore, the source-drain current decreases. As the nanowire diameter increases, a larger potential drop over the GB is needed to overcome the GB barrier, which can result in a decrease of the on-current even when the nanowire diameter increases. The decrease due to multiple GBs of the source-drain current can be phenomenologically viewed as the effect of increasing $|V_T|$ when the total number of GBs is small, but as a combined effect of increasing $|V_T|$ and decreasing the channel effective mobility when the number of GBs is large.

Students from Research

Youngki Yoon, Ph.D. candidate, degree expected, Dec. 2008.
James Fodor, M.S., May 2007.

Publications

(1) Y. Yoon, J. Lin, S. Pearton, and J. Guo, "Role of Grain Boundaries in ZnO Nanowire Transistors," *Journal of Applied Physics*, Vol. 101, 024301, 2007.
(2) S. Pearton, D. Norton, and J. Guo, "ZnO Nanowire Field-Effect Transistors," *IEEE Trans. on Electron Devices*, invited, under review, 2008.

Acknowledgements

We would like to thank Drs. Jenshan Lin, Steve Pearton, Fan Ren, and David Norton for technical discussions, and Canan "Jana" Balaban for administrative assistance.

References

[1] D. C. Look, Mat Sci Eng B-Solid **80** (1-3), 383-387 (2001).
[2] Y. W. Heo, L. C. Tien, Y. Kwon et al., Appl Phys Lett **85** (12), 2274-2276 (2004).
[3] J. Goldberger, D. J. Sirbuly, M. Law et al., J Phys Chem B **109** (1), 9-14 (2005).
[4] D. J. Sirbuly, M. Law, H. Q. Yan et al., J Phys Chem B **109** (32), 15190-15213 (2005).
[5] H. T. Wang, B. S. Kang, F. Ren et al., Appl Phys Lett **86** (24), 243503 (2005).
[6] H. Ohta, M. Orita, M. Hirano et al., J Appl Phys **89** (10), 5720-5725 (2001); R. L. Hoffman, B. J. Norris, and J. F. Wager, Appl Phys Lett **82** (5), 733-735 (2003); A. Tsukazaki, A. Ohtomo, T. Onuma et al., Nat Mater **4** (1), 42-46 (2005); R. L. Hoffman, ZnO: Bulk, Thin Films and Nanostructures, ed. C. Jagadish and S. J. Pearton (2006).
[7] Y. Cui, Z. H. Zhong, D. L. Wang et al., Nano Lett **3** (2), 149-152 (2003).
[8] F. M. Hossain, J. Nishii, S. Takagi et al., Physica E-Low-Dimensional Systems & Nanostructures **21** (2-4), 911-915 (2004).
[9] J. Goldberger, A. I. Hochbaum, R. Fan et al., Nano Lett **6** (5), 973-977 (2006).
[10] J. Wang, E. Polizzi, and M. Lundstrom, J Appl Phys **96** (4), 2192-2203 (2004).
[11] J. Wang, A. Rahman, A. Ghosh et al., IEEE Trans. Electron Dev **52** (7), 1589-1595 (2005).
[12] S. M. Sze, *Physics of semiconductor devices*, 2nd ed. (Wiley, New York, 1981).
[13] J. Guo, J. Wang, E. Polizzi et al., Ieee T Nanotechnol **2** (4), 329-334 (2003).
[14] D. D. D. Ma, C. S. Lee, F. C. K. Au et al., Science **299** (5614), 1874-1877 (2003).

4. Environmentally - Driven Power Source, Power for Wireless Hydrogen Sensor Network - Energy Harvesters

Task PI: Dr. Toshikazu Nishida, Electrical & Computer Engineering, University of Florida
Co-PI: Dr. Khai D. T. Ngo, Electrical & Computer Engineering, University of Florida

Collaborators: Dr. Jenshan Lin, Electrical & Computer Engineering, Dr. Steve Pearton, Material Science & Engineering, Dr. David Norton, Material Science & Engineering, Dr. Fan Ren, Chemical Engineering, Dr. Hugh Fan, Mechanical & Aerospace Engineering, Dr. Mark Sheplak, Mechanical & Aerospace Engineering, University of Florida

Graduate Students: Alex Phipps, Anurag Kasyap, Electrical & Computer Engineering, University of Florida

Research Period: August 3, 2004 to March 31, 2008

Introduction

The goals are to develop a local power source for a self-powered wireless hydrogen sensor network. The multi-energy local power source consisting of energy harvester and power processor harvests vibrational energy for operation during 'dark' conditions and optical (solar) energy for operation during 'light' conditions. The power processor extracts energy from a photovoltaic and a vibration energy harvester and delivers the energy to a reservoir that supports a self-powered hydrogen sensor network. The power source provides power to the wireless transmitter developed by Dr. Lin and the hydrogen sensor developed by Dr. F. Ren.

Background – Energy Harvesting

A typical energy harvesting system can be divided into three functional blocks; the energy harvesting transducer, the power converter, and the load/storage element as shown in Figure 1. The energy harvesting transducer converts ambient waste energy into electricity through one of several transduction mechanisms (thermoelectric, photoelectric, piezoelectric, etc.). The power converter provides matching conditions, including load matching and rectification, between the transducer and the load. The conditioned signal from the power converter is either delivered directly to the load or sent to a battery/capacitor where it can be stored for later use.

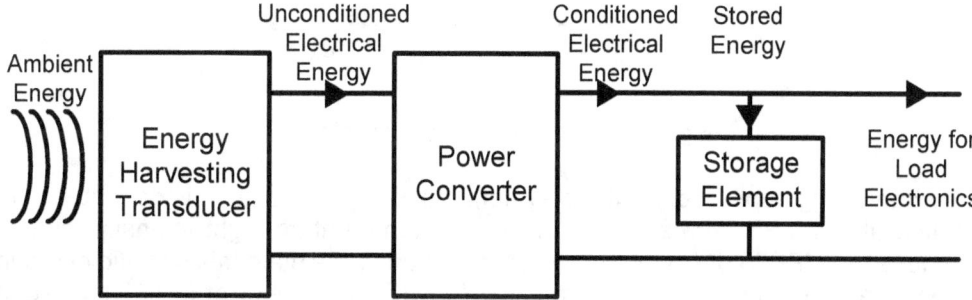

Figure 1: Energy harvesting system.

For this project, two sources for energy harvesting have been considered, solar and vibration energy. Solar energy harvesting uses the photoelectric effect to convert incident light into electrical energy, while vibration energy harvesting relies on the piezoelectric effect to convert strain into electricity. In contrast with unconstrained energy harvesters which have large power output due to their typically large areas or volumes, sensor nodes are generally small to maximize their portability, limiting the amount of harvestable power. The unique features are the development of an energy harvester model that enables optimization to meet a specific power budget such as that required by the wireless hydrogen network and development of power electronics specifically designed to operate at low power.

Background – Solar Energy Harvesting

A solar energy harvesting system allows the sensor node to harvest ambient optical energy during 'light' conditions, and is comprised of the same fundamental blocks as the generic system shown in Figure 1. The energy harvesting transducer is a solar cell, with an I-V characteristic shown in Figure 2. The output power from the solar cell, also shown in Figure 2, is the product of the voltage and current waveforms. When the terminals of the cell are open-circuited, the voltage reaches its open circuit value, V_{oc}, and no current flows. The power delivered in this case is zero. Likewise, when the terminals are short-circuited, the current increases to the short circuit condition, I_{sc}, then there is no voltage difference across the cell. Again the power is zero. Between the short circuit and open circuit cases, there exists a maximum power point (MPP) where the power delivered by the solar cell is at its largest value.

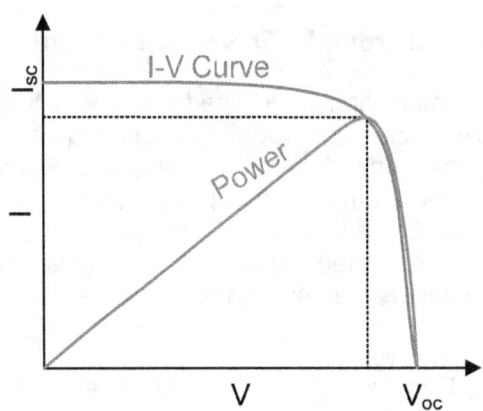

Figure 2: I-V curve for a typical solar cell.

For energy harvesting applications, where system efficiency is crucial, it is desirable to keep the cell operating at the MPP. However, fluctuations in temperature, light intensity, and electrical load can cause the MPP to drift. Control algorithms used to maintain this specific operating point are referred to a maximum power point tracking (MPPT). A number of MPPT techniques have been developed [1,2], but require the use of digital signal processors or complex feedback circuitry. These implementations are not practical for self-powered sensor nodes because of the additional power budget and physical size added by the extra circuitry.

A low power method of MPPT for energy harvesting applications relies on the linear relationship between the open circuit voltage, V_{oc}, and voltage of the MPP, V_{MPP} [1;2]. This relationship can be expressed as

$$V_{MPP} \approx k V_{oc} \tag{0.1}$$

where k is a constant typically between 0.71 and 0.78. Removing the load from the solar cell to measure the open circuit voltage would require extra circuitry (and therefore extra power), so a pilot or reference cell is used instead. The reference cell remains in an open circuit state to provide a reference for the MPPT and must have the same characteristics as the cell providing power. Assuming that both the power cell and the reference cell are small for the self-powered node, the temperature and light differences that they encounter should be negligible.

Using the reference cell technique, a low power MPPT converter has been designed and fabricated for solar energy harvesting. This solar converter, shown in Figure 3, uses two internally generated switching signals, NGate and PGate, to transfer energy to the load. The power converter also maintains the cell voltage at V_{MPP}, with only small perturbations above and below the optimal value. The waveforms from the solar power MPPT converter are presented in Figure 4.

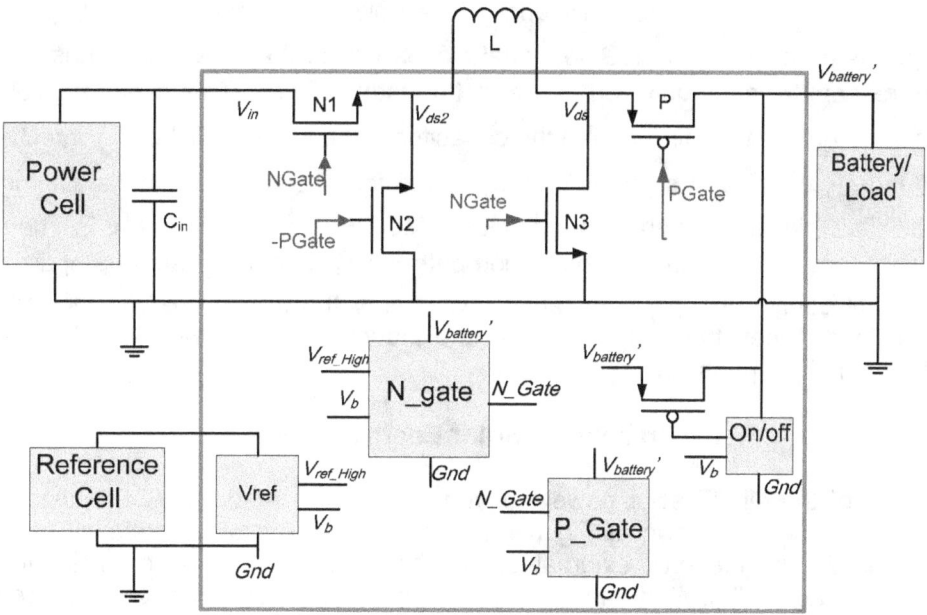

Figure 3: MPPT power converter for solar energy harvesting.

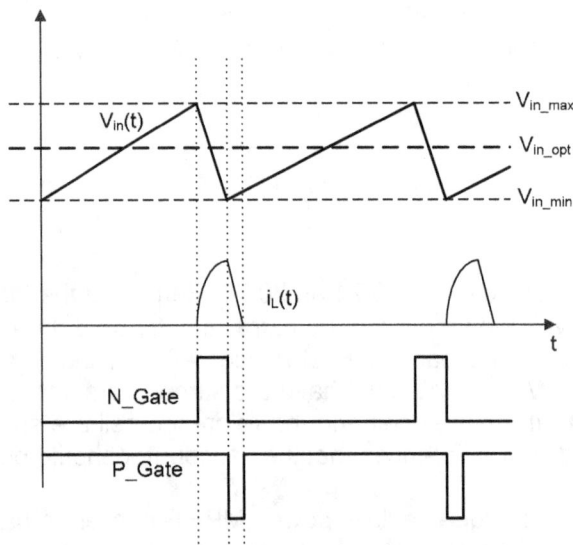

Figure 4: MPPT power converter waveforms.

Operation of the solar converter can be broken down into three phases of operation. During the first phase, the input capacitor, C_{in}, is initially uncharged and all MOSFET switches are open. Current generated by the solar cell flows into C_{in} and the voltage, V_{in}, ramps up to an upper threshold, V_{in_max}. As V_{in} reaches this upper threshold, the second phase begins, and the NGate signal is generated. The NGate signal closes two of the NMOS switches, N1 and N3, forming a resonant circuit between C_{in} and the inductor, L. The capacitor voltage, V_{in}, decreases as energy is transferred from the capacitor to the inductor. When V_{in} reaches a lower threshold, V_{in_min}, the NGate signal is turned off and the second phase ends. During the third phase, the PGate signal is generated which closes MOSFET switches N2 and P. The power cell and the capacitor no longer have a conduction path, so V_{in} begins to ramp up again. Since the inductor current cannot change instantaneously, energy that was stored on the inductor now flows to the load. When the current through the inductor reaches zero, the PGate signal is turned off and the third phase ends.

Experimental – Solar Energy Harvesting

An IC version of the MPPT solar power converter was fabricated by MOSIS using a 0.6 um CMOS process. The IC, shown in Figure 5(a), was designed to generate all of the control signals internally, with the only external components being the solar cells, input capacitor, inductor, and storage cell (battery). Initial testing revealed that the PGate control signal was not being properly generated, but a lack of test pins on the IC prevented a more detailed examination. Fortunately, the design of the IC allowed for the PGate signal to be generated off-chip using the NGate signal as a trigger. The PGate generation circuit, shown in Figure 5(b), uses a monostable multivibrator (HC4538 from Fairchild Semiconductors) to create a variable duration pulse which is triggered off the falling edge of the NGate signal. By adjusting an RC time constant, the duration of the PGate signal can be varied.

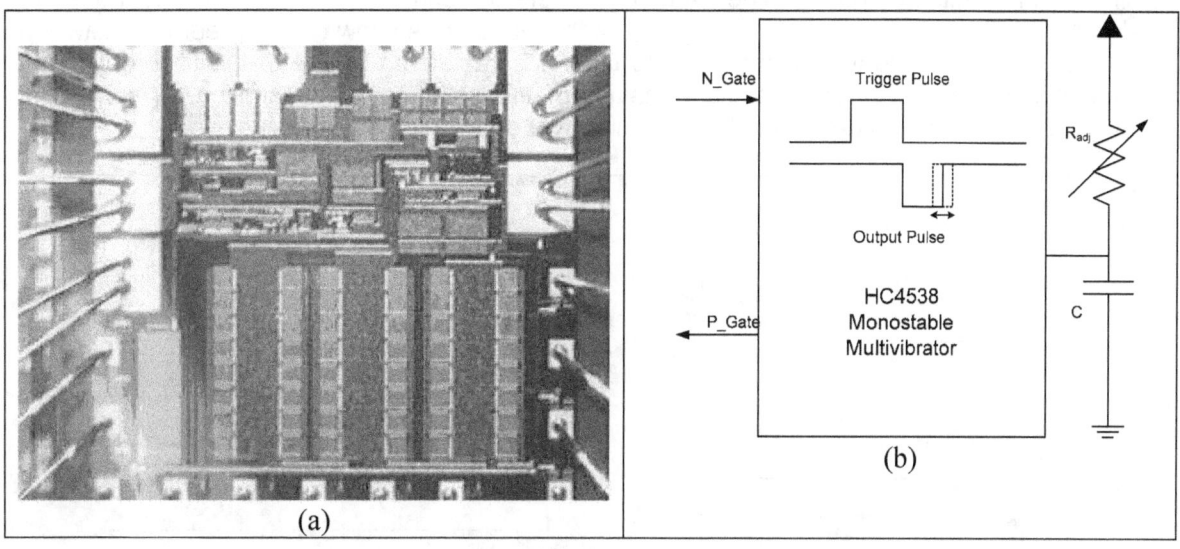

Figure 5: Photograph of (a) IC for the solar energy harvester and (b) schematic of external PGate generation circuit.

A photograph of the experimental setup used to verify the operation of the solar energy harvester can be seen in Figure 6. The solar cells used for these experiments were IXOLAR high efficiency solar cells by IXYS (16.6 mW/cm^2). The capacitor used was a generic 56 nF ceramic and inductor was self-spun to a value of 124 uH. An inductor with relatively large size was used to minimize the parasitic resistance effects. The load for the circuit was a 0.22F high density capacitor, and an adjustable fiber optic light source was used to energize the solar cells.

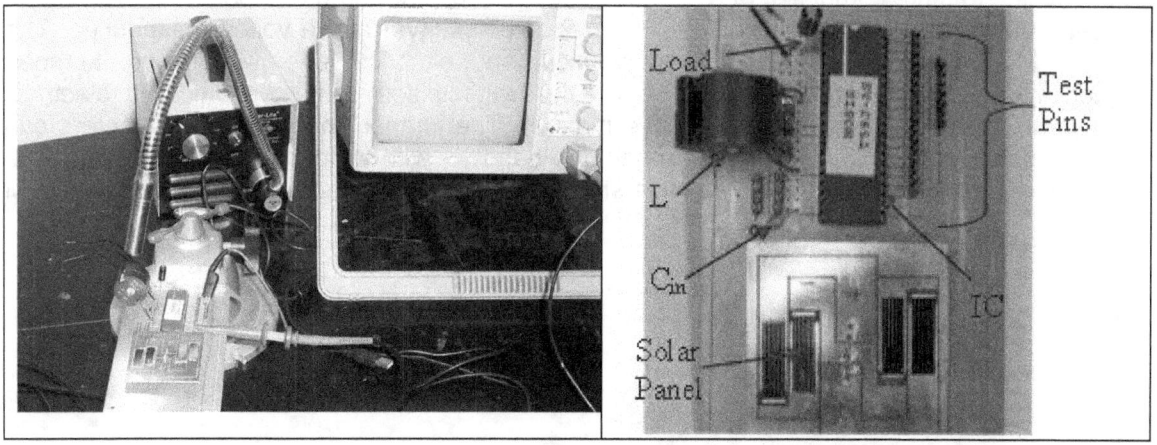

Figure 6: Experimental setup of solar energy harvesting circuit.

Results and Discussion – Solar Energy Harvesting

The V_{in} and NGate waveforms for the solar energy harvester were measured and are compared to simulation results in Figure 7. The simulations were done in Cadence, the CAD tool used to design the IC. The general shapes and timing of the waveforms show good agreement between simulation and experiment. However, due to the uncharacterized nature of the fiber optic light source, it was difficult to capture the behavior of the solar cell in simulation. A solar light simulator may be used to determine the overall power conversion efficiency.

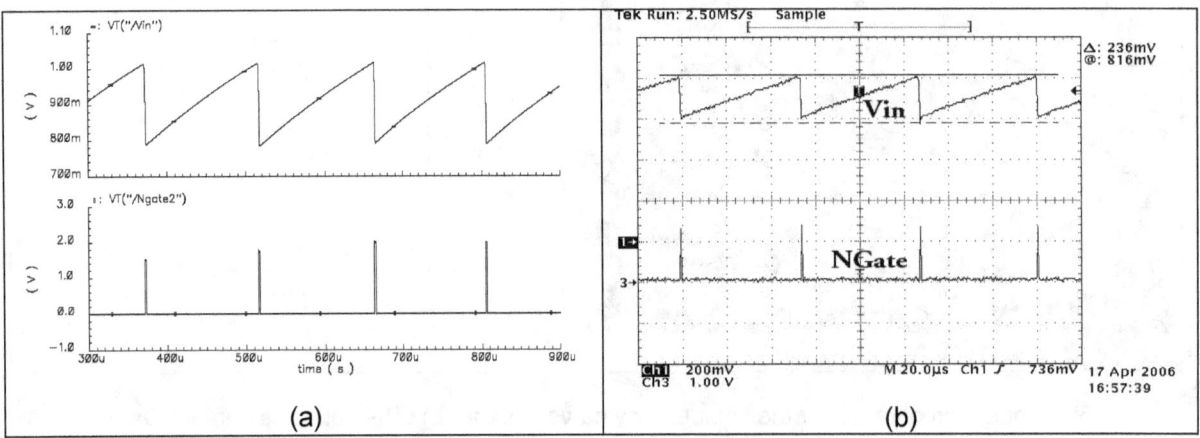

Figure 7: (a) Simulated and (b) experimental waveforms for the solar energy harvester.

Background – Vibration Energy Harvesting

A cantilevered beam configuration is used to harvest ambient vibration energy for the self-powered sensor node during the 'dark' conditions. The cantilever beam, shown in Figure 8(a), is a composite structure comprised of a shim, piezoelectric patch, and proof mass. As the base of the beam is vibrated, strain is applied to the piezoelectric layer and a voltage is induced. The piezoelectric patch thus couples the mechanical and electrical energy domains. Lumped element modeling (LEM) techniques [3] allow the transduction from mechanical to electrical domains to be analyzed in a more intuitive manner. The lumped mechanical parameters can then be added to SPICE simulations to model the behavior with various electronic circuits. The LEM of the piezoelectric cantilever beam is shown in Figure 8(b), and a description of the LEM parameters is given in

Table I.

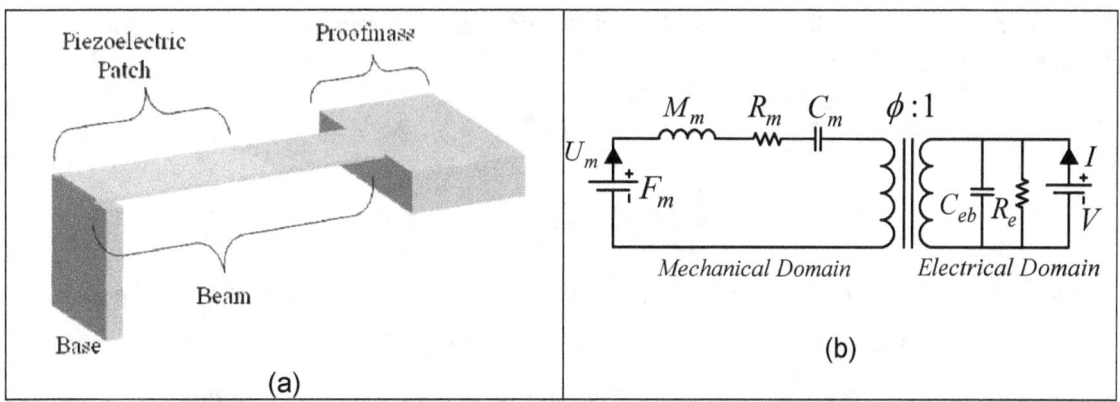

Figure 8: Using LEM techniques, the cantilevered composite beam (a) can be modeled using discrete circuit elements (b).

Table I: LEM parameters.

	Power
F_m	Mechanical Force
U_m	Mechanical Velocity
M_m	Mechanical Mass
C_m	Mechanical Compliance
R_m	Mechanical Damping
R_e	Electrical Damping
C_{eb}	Blocked Electrical Capacitance

A power converter is required in combination with the energy harvester to ensure that maximum power is delivered to the load. Since the voltage and current produced by the beam are sinusoidal, the power converter must also rectify these waveforms to power the sensor electronics and charge storage cells. In order to compare the effectiveness of various power converter topologies, the power extraction efficiency (*PEE*) was defined as a metric for comparison [4]. From power theory, the maximum power transfer, P_{max}, will occur when the energy harvester is loaded with its conjugate matched load. The *PEE* is defined as the ratio of the power extracted from the environment by a specific converter topology to P_{max}.

$$PEE = \frac{P_{converter}}{P_{max}} \times 100 \tag{0.2}$$

The power converter used in this project, the pulsed resonant converter (PRC), has a power extraction efficiency given as

$$PEE_{PRC} = \frac{4}{\pi} \frac{\omega \tau_e}{1+\omega^2 \tau_e^2} \left(1+\exp\left(\frac{-\pi}{\omega \tau_e}\right)\right)^2 \quad (0.3)$$

where the PEE is a function of the variable $\omega \tau$. In (0.3), ω is 2π times the vibration frequency, and τ_e is a characteristic time constant defined as $\tau_e = R_e' C_{eb}$. C_{eb} is the blocked capacitance of the piezoelectric patch, and R_e' represents its internal dissipation. A plot of the PEE vs. $\omega \tau_e$ is shown in Figure 9, which indicates that approximately 75% of the maximum available power could be extracted if the value of $\omega \tau_e$ was properly chosen. Because ω is fixed by the vibration frequency, the optimization variable becomes τ_e, and more specifically its LEM parameters, R_e' and C_{eb}. By proper design of the beam geometry, it should therefore be possible optimize an energy harvester system, employing the PRC, to have a power extraction efficiency of nearly 75%.

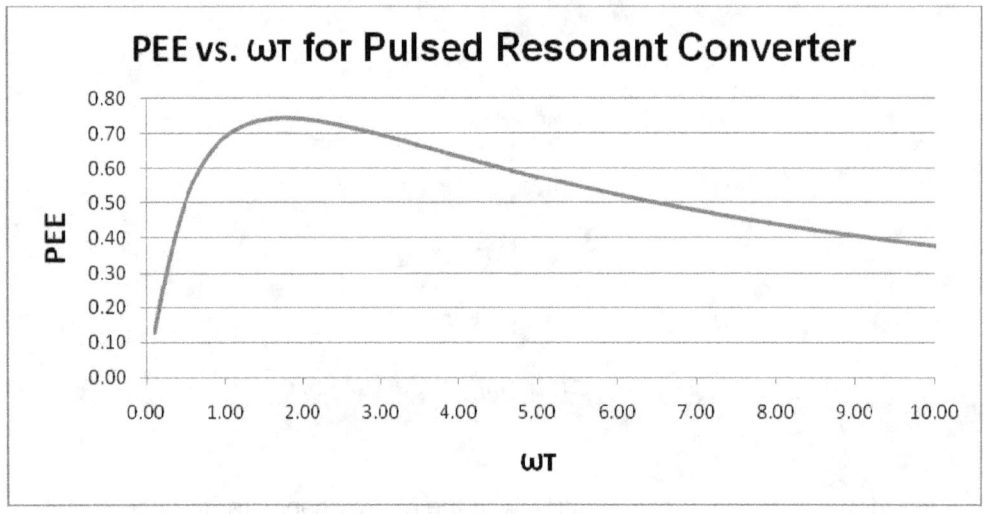

Figure 9: Power extraction efficiency (PEE) vs. $\omega \tau$ for the pulsed resonant converter (PRC).

While the PEE provides an important metric for comparing power converters, efficiency is not the only design goal. For small-scale self-powered systems, total harvested power must also be considered. A trade-off exists between MEMS-scale energy harvesting devices that offer nano-watts of power and higher power meso-scale devices that may exceed the desired sensor node size. Efficiency is somewhat meaningless if the total harvested energy is insufficient to power the node electronics or if the sheer bulk of the energy harvester limits its usefulness.

A new design methodology, *Design for Power*, has been developed to address the trade-offs between power and size. In this approach, the goal is to provide a power budget necessitated by the sensor node while minimizing the total size of the energy harvester. The power converter topology is chosen prior to the beam dimensions, so that the beam can be tailored to the converter. To date, the PRC has been used as the power converter.

Design for power also assumes that the power spectrum of the vibration source is known and is unaffected by the addition of the energy harvester. The resonant frequency of the cantilever beam is designed to coincide with the vibration frequency of the source for a larger response, both mechanical and electrical. When operated at resonance, the harvested power using the PRC is given as

$$P_{PRC} = \frac{\left(F_m^*\right)^2}{2\pi R_e'} \frac{\omega\tau}{1+\omega^2\tau^2}\left(1+e^{\frac{-\pi}{\omega\tau}}\right)^2 \qquad (0.4)$$

where F_m^* represents the mechanical force component reflected into the electrical domain.

Experimental – Vibration Energy Harvesting

For the design for power concept to be applicable, the input power spectral density of the vibration source must be sufficient to allow the energy harvester to meet the power budget. If the required size of the device is too large, the vibration source will be damped out by the harvester. As a test case, an automobile engine was chosen as the vibration source. The acceleration spectrum is shown in 10 for a range of RPM values. Since the engine is relatively large, it provides a vibration source that will not be damped out by meso-scale energy harvesters (on the order of millimeters). The 3000 RPM level was chosen for proof-of-concept design optimization due to both its frequency and acceleration level. Many potential sources for vibration energy harvesting have frequencies between 60 and 250 Hz, and the acceleration level will provide enough energy to operate some low-power electronics.

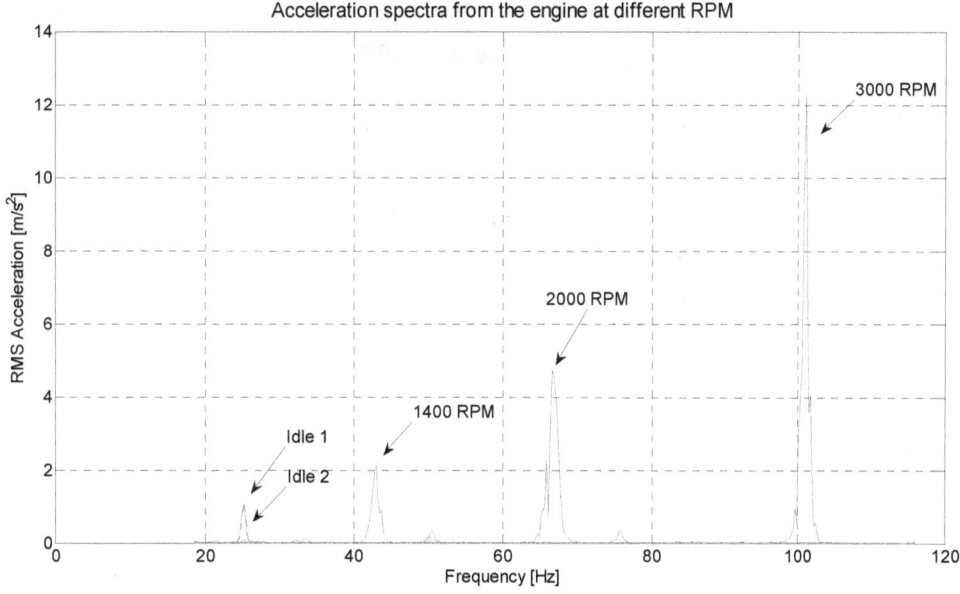

Figure 10: Acceleration spectra of from an automobile engine at different RPM.

Since design for power is basically an optimization of the energy harvester area with input and output power and frequency constraints, a MATLAB code was written to calculate the optimal beam dimensions for a given power budget. The *fmincon* function was used as the optimization algorithm. *fmincon* finds the minimum of a constrained nonlinear, multivariable equation, such as the area of the energy harvester (and its constraints) using a sequential quadratic programming (SQP) method. Several important assumptions were made in the optimizations

- The resonant frequency of the energy harvesting device corresponds to the frequency of the acceleration.
- The energy harvester does not load the source.
- The acceleration applied to the beam does not cause it to leave the linear region of beam operation.
- Parasitic losses of the converter are ignored.
- The piezoelectric patch width equals the shim width.

Results and Discussion

The MATLAB design for power algorithm was run for several different power budgets to gain some insight of its effectiveness. The material parameters used in the optimization are listed in Table II. For these simulations, a Si-PZT beam was designed, with dimensions achievable on a silicon wafer. Micromachining of such beams has been previously demonstrated and good agreement exists between the characterized beams and their models [5]. The results of the optimization for several power budgets are shown in Table III.

Table II: Material parameters used in design for power optimization.

Material Parameters	Silicon	PZT
Young's modulus	169GPa	60GPa
Density	2330 kg/m^3	7500 kg/m^3
d_{31}	N/A	-274 pC/N
Relative permativity	N/A	900

Table III: Design for power optimization results.

DIMENSION	Power>1 uW	Power >60 uW	Power >100 uW
Ls (shim length)	11.7 mm	15.7 mm	16.5 mm
bs (shim width)	960.0 um	1.2 mm	2.0 mm
ts (shim thickness)	120.0 um	150.0 um	150.0 um
Lp (piezo length)	9.6 mm	15.7 mm	16.5 mm
bp (piezo width)	960.0 um	1.2 mm	2.0 mm
tp (piezo thickness)	1.0 um	2.1 um	2.0 um
l (mass length)	11.7 mm	9.3 mm	8.0 mm
w_pm (mass width)	960.0 um	1.5 mm	2.9 mm
H (mass thickness)	1.2 mm	1.5 mm	1.5 mm
frequency	101.0 Hz	100.0 Hz	100.1 Hz
Area	0.22 cm^2	0.39 cm^2	0.72 cm^2

Conclusions

Two forms of energy harvesting have been developed, solar and vibration, for use with self-powered sensor nodes. For solar energy, a low powered MPPT circuit has been designed and fabricated which uses a reference cell to adjust the MPP for changes in temperature, irradiance, and load. Modifications are needed in the IC layout to correct for the PGate signal block, but proper operation was demonstrated when this signal was generated externally.

A new design methodology to improve the efficiency of vibration energy harvesting has also been proposed and simulated. Design for power allows for the optimization of the size of an energy harvesting system to meet a specific power budget. Several of the assumptions made in the design for power method must be further verified and offer future research opportunities. The assumption of resonant operation depends on the mechanical quality factor, Q, of the system. For systems with a very high quality factor, mismatches between the source vibration and the device resonance could cause a high sensitivity of the model to frequency variations. It may be preferential to design the beam with a lower Q so that the sensitivity to frequency is reduced. Parasitic losses should also be considered in a second-order design for power optimization, as these require the system to be designed for a higher power budget.

Interactions

Dr. Rich Waters, Naval Warfare (SPAWAR) Systems Center, San Diego

Dr. Donald Blake, Air Vehicles Directorate, Aeronautical Sciences Division, Wright-Patterson AFB/AFRL

Patents

Nishida, T., Cattafesta, L, Sheplak, M., and Khai D.T. Ngo, "Resonant Energy MEMS Array Processor," U.S. Patent No. 6,954,025, filed 5/13/03, issued October 11, 2005.

Publications

[1] J. Jun, B. Chou, J. Lin, A. Phipps, X. Shengwen, K. Ngo, D. Johnson, A. Kasyap, T. Nishida, H. T. Wang, B. S. Kang, F. Ren, L. C. Tien, P.W. Sadik, D. P. Norton, L. F. Voss, and S. J. Pearton, "A Hydrogen Leakage Detection System Using Self-Powered Wireless Hydrogen Sensor Nodes," Solid-State Electron. 51, pp. 1018-1022, July 2007.

[2] Shengwen Xu, Khai D. T. Ngo, Toshikazu Nishida, Gyo-Bum Chung, and Attma Sharma, "Low Frequency Pulsed Resonant Converter for Energy Harvesting," IEEE Trans. Power Electronics, Vol. 22, pp. 63 – 68, January 2007.

[3] K. Ngo, A. Phipps, T. Nishida, J. Lin, S. Xu, "Power Converters for Piezoelectric Energy Extraction," presented at ASME International Mechanical Engineering Congress and Exposition, Chicago, IL, November 5-10, 2006.

[4] Kasyap, A. Phipps, M. Sheplak, K. Ngo, T. Nishida, L. Cattafesta, "Lumped Element Modeling of Piezoelectric Cantilever Beams for Vibrational Energy Reclamation," presented at ASME International Mechanical Engineering Congress and Exposition, Chicago, IL, November 5-10, 2006.

[5] T. Nishida, J. Lin, K. Ngo, F. Ren, D. Norton, S. Pearton, L. Cattafesta, M. Sheplak, J. Jun, A. Kasyap, D. Johnson, A Phipps, "Wireless Hydrogen Sensor Self-Powered Using Ambient Vibration and Light," presented at ASME International Mechanical Engineering Congress and Exposition, Chicago, IL, November 5-10, 2006.

[6] J. Jun, B. Chou, J. Lin, A. Phipps, X. Shengwen, K. Ngo, D. Johnson, A. Kasyap, T. Nishida, H. T. Wang, B. S. Kang, F. Ren, L. C. Tien, P. W. Sadik, D. P. Norton, L. F. Voss, and S. J. Pearton, "Low-Power Detection of Hydrogen Leakage Using a Self-Powered Wireless Hydrogen Sensor Node," 2006 Spring Meeting of American Inst. Chem. Engineers, Orlando, FL, April 23-27, 2006.

[7] Shengwen Xu, Khai D. T. Ngo, Toshikazu Nishida, Gyo-Bum Chung, and Attma Sharma, "Converter and Controller for Micro-Power Energy Harvesting," Proceedings of IEEE Applied Power Electronics Conference, 2005, pp. 226-230.

Students from Research

Anurag Kasyap (Graduated 2007, PhD Mechanical and Aerospace Engineering, Employed at AdaptivEnergy)
Shengwen Xu (Employed at Intel)
Alex Phipps (PhD student)

Acknowledgements

We would like to thank all the collaborators for technical discussions.

Bibliography

[1] J. H. R. Enslin, M. S. Wolf, D. B. Snyman, and W. Swiegers, "Integrated photovoltaic maximum power point tracking converter," *Ieee Transactions on Industrial Electronics*, vol. 44, no. 6, pp. 769-773, Dec.1997.

[2] G. W. Hart, H. M. Branz, and C. H. Cox, "Experimental Tests of Open-Loop Maximum-Power-Point Tracking Techniques for Photovoltaic Arrays," *Solar Cells*, vol. 13, no. 2, pp. 185-195, 1984.

[3] Stephen D.Senturia, *Microsystem Design*. New York: Kluwar Academic Publishers, 2001.

[4] Khai D.T.Ngo, Alex Phipps, Toshikazu Nishida, Jenshan Lin, and Shengwen Xu, "Power Converters for Piezoelectric Energy Extraction," in *ASME IMECE* 2006.

[5] Anarag Kasyap, "Development of MEMS-based piezoelectric cantilever arrays for vibrational energy harvesting." Ph.D Aerospace Engineering, University of Florida, 2007.

[6] Fei Liu, Alex Phipps, Stephen Horowitz, Louis Cattafesta, Toshikazu Nishida, and Mark Sheplak, "Acoustic energy harvesting using an electromechanical Helmholtz resonator," *The Journal of the Acoustical Society of America*, 2008.

[7] J. Chen and K. D. T. Ngo, "Alternate forms of the PWM switch model in discontinuous conduction mode [DC-DC converters]," *Aerospace and Electronic Systems, IEEE Transactions on*, vol. 37, no. 2, pp. 754-758, 2001.

5. Detecting Hydrogen by Enzyme-Catalyzed Electrochemistry

Task PI: Dr. Z. Hugh Fan, Mechanical & Aerospace Engineering, University of Florida

Collaborator: Dr. Jenshan Lin, Electrical & Computer Engineering, University of Florida

Research Period: August 3, 2004 to March 31, 2008

Abstract

This project was to develop a novel hydrogen sensor using an enzyme-catalyzed reaction and microfluidic technology. The potential benefits over the state-of-the-art palladium- or its alloy-based hydrogen sensors include the ability to operate at ambient temperatures, the ability to operate in background gases due to the specificity of the enzyme and fast response time. During the funding period, we demonstrated the hydrogen detection scheme of using hydrogenase to accomplish ambient temperature detection. We reported detection of H_2 ranging from 1% to 100%. We also initiated the work to adapt this enzyme-based detection scheme to a microfluidic device, which is an ideal platform to carry out the catalytical reactions prior to electrochemical detection. In addition, we developed a technique to fabricate microelectrodes by using airbrushing, which costs much less than screen-printing. We found that these microelectrodes satisfactory for electrochemical detection.

Introduction

One of the future alternatives to current fossil-based transportation fuels has been centered on hydrogen (H_2), even though there is a lot of debate on the subject.[1] Currently, H_2 is the primary energy source of today's space exploration projects (e.g., as rocket propellant). It is also used in fuel cells that power a variety of machinery including automobiles. However, H_2 is a colorless, odorless, and flammable gas with a lower explosive limit of 4% in air. Therefore reliable H_2 sensors are required to detect possible leaks wherever H_2 is produced, stored, or used.

To detect H_2, Lundström et al. pioneered a sensor that consists of a palladium Schottky diode on a silicon substrate.[2,3] The sensing device consists of a hydrogen-sensitive metal (palladium or Pd alloy) deposited on a metal-oxide-semiconductor (MOS) structure. Further development[4-7] of this type of hydrogen sensors has led to commercialization and its use to detect H_2 leaks during pre-launches of space vehicles.[8] Recent advances in H_2 sensor technology include those using a gallium nitride MOS diode,[9] Pd-covered silicon microchannels,[10] Pd-coated micromachined cantilevers,[11] and silver/palladium nanowires.[12] One of the concerns for the sensors using palladium or the like is the requirement of a high operating temperature (>200 °C) and further elevated temperatures (>500 °C) to reactivate the sensing element.[13] To address these concerns, this project was to investigate a H_2 sensing scheme based on enzyme-catalyzed electrochemical detection. The principle of using a biological assay for detecting hydrogen is shown as follows:

$$H_2 + 2BV^{2+} \xrightarrow{\text{hydrogenase}} 2H^+ + 2BV^+$$

The enzyme, hydrogenase, catalyzes the oxidation of H_2; BV^{2+} (benzyl viologen) acts as the electron acceptor, producing BV^+ that can be detected electrochemically. The expected benefits include the ability to operate at an ambient temperature and enhanced selectivity due to the specificity of the enzyme.

Experimental

Reagents and Materials. Potassium chloride (enzyme-grade), potassium phosphate monobasic, and β-nicotinamide adenine dinucleotide (NADH) disodium salt hydrate were purchased from Fisher Scientific (Atlanta, GA). Benzyl viologen (BV^{2+}) dichloride was from Sigma-Aldrich (St. Louis, MO). Hydrogen (99.99% pure), nitrogen (industrial grade), and oxygen (industrial grade) cylinders were obtained from Strate Welding Supply Company (Gainesville, FL). Potassium ferrocyanide trihydrate was obtained from Acros Organics (Belgium) and carbon ink paste (BQ225) was from DuPont (Research Triangle Park, NC). Topas® 8007 films were obtained from Topas Advanced Polymers, Inc. (formerly Ticona of Celanese, Florence, KY) while Zeonor® 1020 resins were from Zeon Chemicals Co. (Louisville, KY).

All solutions were prepared using water purified from Barnstead Nanopure Water System (Model: D11911, Dubuque, Iowa). Enzymes were prepared in an aqueous buffer (either phosphate pH 7.0 or Tris-HCl, pH 8.0) consisting of 0.1 M KCl. All experiments were carried out using electrolyte solution containing 0.1 mg/ml hydrogenase, 20 μM NADH (cofactor), 1 mM BV^{2+}, 0.1 M KCl (as the supporting electrolyte), and 0.02 M potassium phosphate, pH 8.0. The enzyme was incubated with NADH for 5 minutes before exposure to H_2 in order to reactivate the enzyme.[14] Solutions were freshly prepared and degassed with nitrogen for each experiment.

Electrochemical Detection. Detection of H_2 was carried out by exposing the gas sample to 1 ml of an aqueous enzyme solution in a flow cell with a height of 4 cm and a diameter of 2.5 cm. Full interaction between the gas and aqueous phases was ensured by bubbling the gas sample through the stirred solution via a pipette tip. The flow cell was sealed at the top via a custom-made removable aluminum cap and o-rings. In the cap, five ports allowed for access of gas inlet/outlet and three electrodes for electrochemical detection. Hydrogen was oxidized by hydrogenase and BV^{2+}, and the product, BV^+, was then detected via chronoamperometry (CA). Gas agitation and stirring were off when CA was performed. To decrease signal variation in detection, the flow cell was kept at 30 °C.

Chronoamperometry was carried out using a polished, flat disc gold working electrode (diameter of 2 mm), a platinum wire counter electrode, and an Ag/AgCl (saturated KCl) reference electrode, all of which were obtained from CH Instruments, Inc. (Austin, TX). All three electrodes were placed equidistant from each other in the solution. The voltage control and current measurement were performed using CH Instruments' 600B electrochemical analyzer and the corresponding software run on an MS Windows-based desktop computer. Single-step chronoamperometry at a potential of 0.05 V against the reference electrode was performed; data acquisition was at a rate of 1.0 kHz.

Microelectrode Fabrication. The pattern of microelectrodes is designed using AutoCAD, as shown in **Figure 1a**. The electrodes are 300 μm wide and 4.6 mm long while the electronic contact pads are 2 mm x 2 mm. The CAD (computer-aided design) file was sent to a vendor (Global Stencil, Dallas, TX) who created a 0.012-inch-thick (1 inch = 25.4 mm) stainless steel stencil by cutting the patterned areas using a laser. The stencil was placed against a 25 mm x 75 mm Topas film and the assembly was ready for airbrushing.

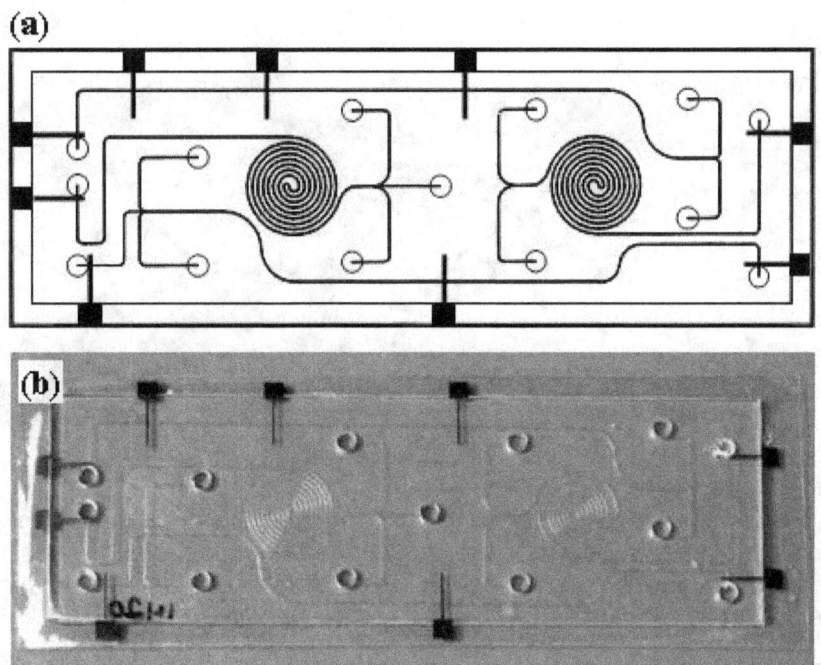

Figure 1. (a) Layout of microfluidic devices used in this work. The size of the device is 25 mm x 75 mm. The device consists of 4 different designs, each of which has multiple inlets, a long channel for reactions, ink electrodes for electrochemical detection, and one outlet. Each black square on the device edge represents the electronic contact pad, and it connects with an electrode indicated by a thick trace. *(b)* Picture of a plastic microfluidic device for H_2 detection. Channels and wells are in the plastic substrate, which was laminated with a thin film patterned with ink electrodes. Electrodes were used for electrochemical detection.

Microelectrodes were fabricated by using airbrushing. Ink solutions were prepared by diluting carbon ink paste using acetone to obtain a concentration ranging from 0.08 g/mL to 0.4 g/mL (weight of ink/volume of acetone). After homogenizing the mixture for 10 seconds using a vortex mixer (Fisher Scientific), 70 μl of the ink solution was dispensed using a pipette into the ink cup of an airbrush (Badger 200SG, Franklin Park, IL). As shown in **Figure 2**, the nozzle of the airbrush was opened by the control screw at the back (about one and three quarters of rotation). Compressed air was introduced into the airbrush by pressing the trigger. The ink solution was sprayed in a smooth stream at the openings of the stencil onto the plastic film. Except where specified otherwise, operation conditions are as follows. The scan time of each spray was 1-2 seconds. After one minute dry in the air, a second spray was carried out at the same location. The airbrush pressure was set at 20 psi (1 pound per inch square = 6.89 kPa). The resultant ink electrodes were air dried for 5 minutes prior to placing them in a furnace. The curing process took place at 70 °C for 1 hour, and the resultant ink electrodes adhered to the plastic film very well. The width and thickness of ink electrodes were measured using a Dektak II profilometer (Veeco, Woodbury, NY).

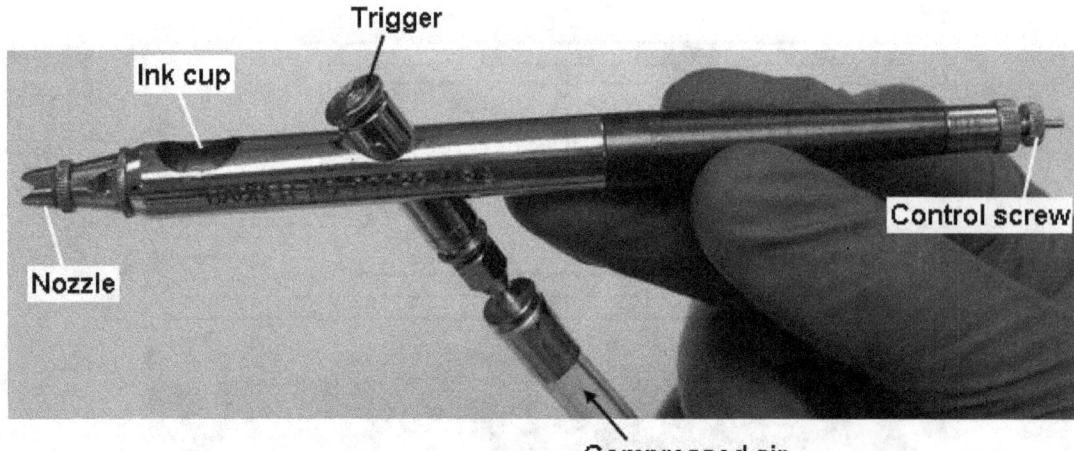

Figure 2. Picture of an airbrush showing the components and operation. Ink was filled in the cup, fed into the stream due to gravity and the Venturi effect. Compressed air was introduced into the airbrush by pressing the trigger. The mixing ratio of ink and air at the nozzle was controlled by the screw at the back of the handle.

Device Fabrication. Plastic microfluidic devices were fabricated using the procedure described previously[15] with minor modification. In brief, a photomask was designed using AutoCAD. The pattern on the photomask was then reproduced in a glass plate via photolithography. Electroplating on the glass plate generated a nickel mold, which was employed to produce plastic parts from Zeonor resins using a hydraulic press (Carver, Wabash, IN). The molding temperature and pressure were set at 146 °C and 5000 psi, respectively. Each plastic part was trimmed into a 25 mm x 75 mm substrate using a CNC milling machine, which was also used to drilled holes (2 mm diameter) at the ends of channels. After cleaning, the plastic substrate was aligned with a 0.1-mm thick Topas film patterned with ink electrodes. Efforts were made to align the electrodes with the end of a channel, but not in the wells, for better electrochemical detection. The assembled substrate and film were briefly preheated at 95 °C and then laminated at 110 °C using a thermal laminator (Catena 35, GBC, Northbrook, IL) while the rollers of the laminator was set at 1/8 inch apart. **Figure 1b** shows a picture of the plastic device used in this work. The thickness of the device is 1.5 mm. Channels are 40 μm deep and 110 μm wide, and its hydraulic diameter[16] is 56 μm. The dimensions of channels were measured using the Dektak II surface profiler before lamination.

Electrochemical Detection Using Microelectrodes. To study the ink electrodes, initial tests were carried out without laminating them into a device for convenience. The ink electrode was simply placed in a 2 ml of electrolytes containing 0.5 mM ferrocyanide and 0.1 M KCl in a 15 ml vial. The half length of the ink electrode was submerged in the solution while a platinum wire was used as the counter electrode and an Ag/AgCl electrode as the reference electrode. Single-step chronoamperometry for ferrocyanide was operated at a potential of 0.5 V against the reference electrode; data acquisition was at a rate of 1.0 kHz.

For ink electrodes laminated in a microfluidic device, electrolytes were introduced into channels of the device using syringe pumps from World Precision Instruments (Sarasota, FL). The pumps were controlled by a controller that was able to command four pumps simultaneously. Fluids were delivered at 200 μL/min through a tube that was connected to the device via 1/16 inch flangeless ferrules (Upchurch, Oak Harbor, WA) that were glued to the wells by epoxy. The device was placed on the sample stage of an inverted microscope (IX51, Olympus America Inc,

Melville, NY) for monitoring flows. For electrochemical detection, the ink electrode at the end of the channel was the working electrode while the counter electrode was placed in the well; two-electrode configuration was used for electrochemical detection. The counter electrode was placed in a way that it did not touch the bottom of the well and it also functioned as a pseudo-reference electrode.[17]

Results and Discussion

Electrochemical Detection. As detailed in the Experimental Section, H_2 gas samples are prepared by pre-mixing H_2 with N_2 and/or O_2 at a pre-determined ratio. Each sample was introduced into the flow cell where electrochemical detection took place. The flow cell contained hydrogenase, NADH as a cofactor, and BV^{2+} as an electron acceptor. Oxidation of H_2 was catalyzed by hydrogenase and the released electrons were accepted by BV^{2+}, which was then reduced to BV^+. The accumulated amount of BV^+ after a certain period of time was detected using chronoamperometry and the measured current was used to correlate with the amount of H_2.

The current-time curves from a potential step in CA are shown in **Figure 3**. Two sets of data are included; one for 1% H_2 and another for 5% H_2 in the gas mixture. We denoted oxidation current as positive in the figures, so that larger signals corresponded to higher concentrations of H_2. For each H_2 concentration, four experimental trials were performed and the variations were primarily due to difference in gas delivery. We found that timing during all steps of the detection protocol, from enzyme activation to final signal measurement, was critical in producing consistent results.

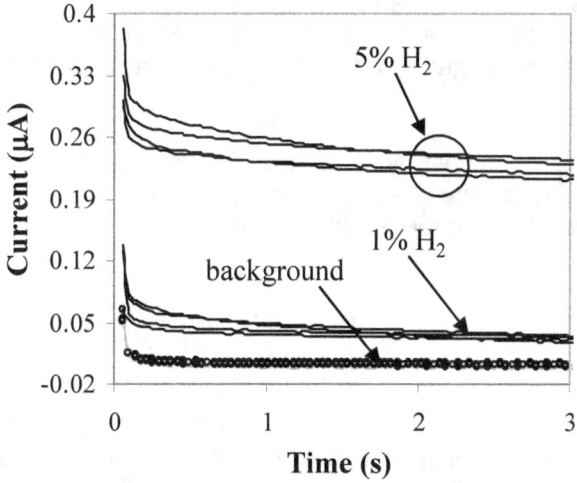

Figure 3. Current-time curves obtained in chronoamperometry for 1 and 5% H_2 gas mixtures. Four repeats are shown for each concentration, along with corresponding background measurements. Background data points are labeled with circles for clarity and exhibit little difference between trials. The experiments were carried out in a solution containing 0.1 mg/ml hydrogenase, 20 μM NADH, 1 mM BV^{2+}, 0.1 M KCl, and 0.02 M potassium phosphate, pH 8.0.

According to the chronoamperometry theory,[18] the current, i, changes with time, t, according to the Cottrell equation: $i = \dfrac{nFAD^{1/2}C}{\pi^{1/2}t^{1/2}}$, where n is the number of electrons transferred per molecule, F is Faraday's constant (96,500 C/mol), A is electrode area, D is diffusion coefficient, and C is the concentration of redox molecules. As a result, a plot of current versus the reciprocal of the square root of time ($t^{-1/2}$) should be linear. We plotted the data of 5% H_2 in Figure 3 and obtained straight lines, indicating that experimental results agreed with the theory of chronoamperometry.

To eliminate background signals, we performed two CA measurements for every detection trial. First, the background signal of the electrolyte solution was measured after the solution had been purged with N_2 for 4 minutes. Next the signal of the same solution was determined after its exposure to a gas sample for a period of time. The exposure time depended on the desired intensity of signal response and decreased with increasing H_2 concentration. For the range of 1% to 10% H_2 the exposure time was fixed at 6 minutes. The net signal was determined as the difference in current (averaged over 10 milliseconds) between the background and signal measurements at 3 seconds. The net signal was proportional to the difference in the analyte (BV^+) concentrations between the H_2-exposed and blank solution.

To simplify signal measurement, the background signal was maintained close to 0 A by using an applied voltage at the working electrode equal to the open circuit potential (OCP). OCP is the potential at which no current flows in the cell, and it was measured by the CHI600B analyzer.[19] This arrangement ensured that all current in the signal measurement was caused by changes in the solution taking place during gas sample exposure. OCP ranged from 0.05 to -0.065 V, and thus the applied potential was set at 0.05 V, well above the redox potential of BV^+ (E_0 = -0.550 V vs. Ag/AgCl in saturated KCl), and resulted in background measurements in the order of 1×10^{-9} A. Figure 3 shows that a significant signal was obtained at 1% H_2 gas concentration; and current increased when H_2 concentration changed from 1% to 5%. Variation in signal increased with H_2 concentration but, for a given net signal, consistently represented ~7% of the mean value.

Enzyme Kinetics. In order to assess enzyme activity and overall performance of the detection protocol, kinetics of the system were analyzed according to the Michaelis-Menten equation.[20] The equation can be converted into the equation below,

$$\frac{1}{V} = \frac{1}{V_{max}} + \frac{K_m}{V_{max}} \frac{1}{[S]}$$

where V is the rate of formation of the product, V_{max} is the maximum rate of reaction at a given enzyme concentration, K_m is the Michaelis constant, and S is the concentration of the substrate. This equation was used to obtain the Lineweaver-Burke plot.[20] A plot of $1/V$ versus of $1/[S]$ fits a straight line, which can be used to determine V_{max} from the intercept and K_m from the slope. We used a series of substrate concentration (i.e., dissolved H_2 in solution) to obtain the corresponding rates of the enzymatic reaction. The concentration of dissolved H_2 in solution was estimated using Henry's law.[21]

Figure 4 shows the Lineweaver-Burke plot obtained from experimental results. The trend line has a correlation coefficient (R^2) of 0.9976 and error is approximately 5% of the mean value at each data point (based on four repeats). Linear regression resulted in a slope and intercept, from which we calculated 4.90×10^{-5} M/s and 3.09×10^{-3} M for V_{max} and K_m, respectively. This result indicates that the H_2-BV^{2+} assay at low H_2 concentration indeed follows the Michaelis-Menten kinetics.

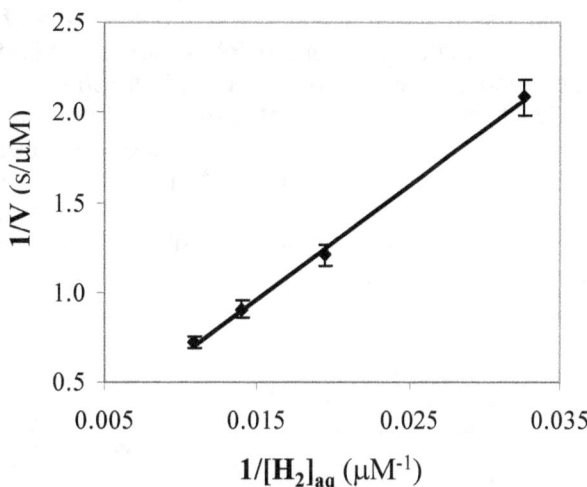

Figure 4. Lineweaver-Burke plot of the reciprocal of the reaction rate as a function of the reciprocal of the substrate concentration. The concentration of H_2 dissolved in solution was calculated from 3, 5, 7 and 9% H_2 gas mixtures, whereas the reaction rate was calculated from the rate of formation of [BV^+], which was obtained from CA.

Using V_{max} obtained from the plot, the H_2-BV^{2+} activity of the SH was determined to be 0.9 U/mg at 9% H_2, which is ~4% of the value (24 U/mg) reported for SH batch A with the H_2-NAD^+ assay.[14] Under optimal assay conditions the H_2-BV^{2+} activity is about 40% of the H_2-NAD^+ activity.[14] The electrochemical detection, however, required a different experimental setup. Buffer composition, enzyme concentration, and reactivation procedure had to be modified, resulting in decreased enzymatic activity.

Table 1 summarizes the results of the kinetic investigation using three different batches of SH enzyme. The H_2 oxidizing activities of all samples were in the same range, both with NAD^+ and BV^{2+} as electron acceptors. The K_m values for H_2 in the H_2-BV^{2+} assay and the resulting currents were also comparable. Thus our H_2 sensing system yielded reproducible results.

Table 1. Comparison of kinetic data among three batches of enzymes.

SH[1]	V_{max} (µM/s)	K_m (mM)	H_2-BV^+ activity[2] (µmol H_2/min)	H_2-NAD^+ activity[3] (µmol H_2/min)
A	49.0	3.09	0.88	24.4
B	71.7	3.01	1.3	31.2
C	61.2	3.26	1.0	32.6

Note: 1. Three batches of enzymes were prepared differently.
2. The values were determined from V_{max} at 9% H_2. The experimental setup contained 0.1 mg/ml hydrogenase, 20 µM NADH, 1 mM BV^{2+}, 0.1 M KCl, and 0.02 M potassium phosphate, pH 8.0.
3. The values were determined under H_2 saturation in 50 mM Tris-HCl buffer pH 8.0 without prior reactivation with NADH.

Calibration Curve and O_2 Sensitivity of the H_2 sensing element. We obtained a calibration curve showing the linear response between the signal and percentage of H_2 in gas samples, as shown in **Figure 5**. The correlation coefficient of the linear regression line (R^2) is 0.9951 with an average of 5.0% variation at each concentration (based on four repeats). H_2 concentration down to 1% could be detected. As discussed above, the exposure time was fixed at 6 minutes for 1-9% H_2. At higher concentrations of concentrations, less exposure times (1 minute for 25-50% and 30 seconds for 75-100% H_2) were sufficient to achieve easily detectable signals. The difference in the exposure time was corrected by normalizing the current against the exposure time. One point worthy of note is that careful management of the total gas exposure time was more important for experiments using lower exposure times, since a small deviation led to larger variation in signal response.

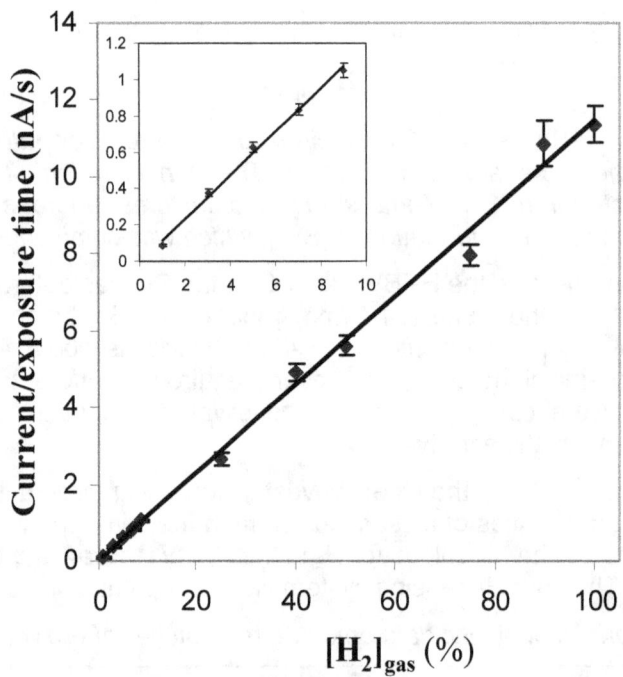

Figure 5. Calibration curves between current per unit of exposure time and percentage of H_2 in gas mixtures. The current was obtained by chronoamperometric detection of BV^+, which was the reaction product of BV^{2+} and H_2 catalyzed by hydrogenase. The gas exposure time was fixed to 6 minutes for 1-9%, 1 minute for 25-50%, and 0.5 minutes for 75-100% H_2; the difference in the exposure time was corrected by normalizing the current against the exposure time. The gas mixture contained a pre-determined ratio of H_2 in nitrogen. The insert shows the calibration curve for the range of 1-9% H_2.

Since hydrogenases are generally known to be sensitive to O_2; exposure to O_2 could lead to temporary and permanent deactivation of the enzyme.[22] In contrast to the structurally well characterized standard Ni-Fe hydrogenases, the SH of *R. eutropha* has high tolerance to O_2.[23] Reactivation with NADH removes an oxygen species bound to the active site. O_2 tolerance is caused by an extra cyanide ligated to the Ni. Removal of the extra cyanide results in lower catalytic activity and an irreversible inactivation of the SH.[23] In order to determine how O_2 affects the present H_2 sensing setup, we studied SH exposure to gas samples containing 20% O_2 (mixed with N_2 and H_2). **Table 2** shows that inclusion of O_2 has a consistent negative effect on net detection signal, causing an average decrease in signal of 19%. However, a linear calibration curve was obtained with the correlation coefficient (R^2) of 0.9918 from the data in the O_2 environment. All results were averaged from four trials at each H_2 and O_2 concentration. The decrease in the signal was due to the fact that the BV^+ produced by the enzymatic reaction was immediately oxidized by O_2. As a result, this effect can be eliminated by choosing another electron acceptor (other than BV^{2+}) that is not affected by oxygen. Therefore, we confirmed that SH is resistant to O_2 deactivation and that SH was able to function as a H_2 sensing element in the presence of O_2.

Table 2. Effects of oxygen on hydrogenase-catalyzed H_2 detection.[1]

H_2% (gas)	$[H_2]_{aq}$ (μM)	Current (μA) no O_2	Current (μA) 20% O_2	Current decrease (%)
1	10.3	0.031	0.024	23.2
3	30.8	0.135	0.111	14.6
5	51.3	0.225	0.193	16.0
7	71.8	0.301	0.256	22.0
9	92.3	0.378	0.314	19.2

Note: 1. The experiments were carried out in a solution containing 0.1 mg/ml hydrogenase, 20 μM NADH, 1 mM BV^{2+}, 0.1 M KCl, and 0.02 M potassium phosphate, pH 8.0.

Microelectrode Fabrication. Screen-printing is a printing technique that is traditionally used in fine arts and in commercial printing.[24] It has been exploited as a relatively simple and versatile technique to fabricate electrodes for electrochemical analysis.[25-30] Airbrushing is another technique to produce a variety of effects for advertising art, photo retouching, and renderings in technical and commercial art fields.[31] We looked into the possibility of using airbrushing for fabricating microelectrodes for electrochemical detection.

The airbrush in **Figure 1** mixes air and ink, producing a thoroughly atomized fine dot spray pattern. It has a top-mounted cup in which gravity helps draw ink into the airbrush, requiring less air pressure to operate. Other types of airbrushes, including external mix and bottom or side feed, also exist; the details can be obtained from manufacturers or in the literature.[31]

We found that airbrushing allowed fabricating well defined electrodes with ease. It also required a smaller amount of ink than screen-printing. Picture of airbrushed ink electrodes integrated in a microfluidic device is shown in **Figure 2**. We studied the possible effects of the airbrush operation parameters, including the ink concentration, the number of spray repetitions and the elapsed time of each spray, the curing time, and the airbrush pressure, on the electrode properties. Each operation parameter was studied by leaving all the other variables constant. The physical integrity of each electrode was examined under a microscope. The width and depth of electrodes were measured by profilometry. Chronoamperometry and voltammetry of ferrocyanide were performed to study the reproducibility of electrode fabrication under one condition.

As explained in the *Experimental Section*, the carbon ink paste was diluted with acetone in order to be properly sprayed through the airbrush. The concentration of the resultant ink ranged from 0.08 to 0.4 g/ml. We found that no significant dependence existed between the ink concentration and the thickness or width of the resultant ink electrodes, as shown in **Figure 6**. It was expected that the width of the ink electrodes was defined by the opening of the stencil; the average of all results were close to the designed value of 300 μm. The standard deviations are indicated by the error bars, and they resulted from the variations among different openings in the stencil and the experimental variations. For electrochemical detection, the electrode width is more important than the thickness, as it directly relates to the electrode area and thus the current. We then chose 0.24 g/ml as the ink concentration because of its smaller variation, ease of operation, and physical characteristics of the resultant electrodes. Lower concentrations of the ink could be easily sprayed, but they tended to bleed and smear underneath the stencil. This partially explains the relatively large deviation when 0.08 g/ml ink was used. The higher concentrations of the ink were difficult to spray due to its high viscosity and the small opening of the airbrush nozzle.

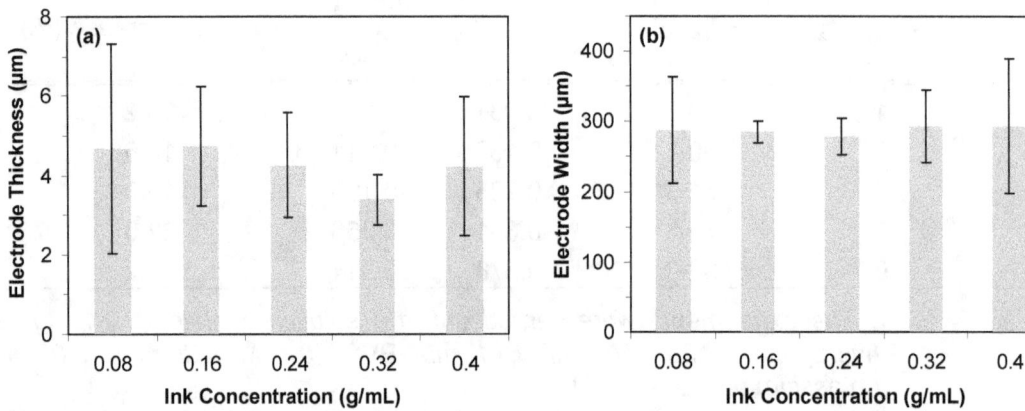

Figure 6. The effect of the ink concentration on the thickness *(a)* and width *(b)* of microelectrodes fabricated by airbrushing. Both width and thickness were measured by a profilometer. The error bars indicate the standard deviations obtained from three to six electrodes.

We also studied the effect of other operation parameters, including the number of spray repetitions at 2, 3, and 4; the duration of each spray at 1, 2, and 3 seconds; the curing time at 1 and 2 hours, and the compressed air pressure at 10, 20, and 30 psi. We did not observe significant difference in electrode physical appearance, electrode dimension, and electrochemical detection. As a result, we chose the operation parameters from the angle of ease of operation; these parameters are listed in **Table 3**.

Table 3. Airbrushing operation parameters for fabricating ink electrodes

Operation Parameters	Optimum Values
Ink concentration	0.24 g/mL
Number of sprays	2
Spray time	1-2 s
Airbrush pressure	20 psi
Curing temperature and time	70 °C for 1 hr

Microelectrode Cyclic Voltammetry. Electrochemical property of airbrushed ink electrodes was studied by running cyclic voltammetry. The experiments were performed by using an ink electrode as the working electrode, which was connected to the CHI electroanalyzer through an alligator clip clamped on the contact pad of the airbrushed electrode. The ink electrode was about 2.3 mm immersed into a ferrocyanide solution. An Ag/AgCl electrode was used as the reference electrode and a platinum wire functioned as the counter electrode. A cyclic voltammogram (CV) of ferrocyanide (a model redox molecule) using airbrushed ink electrode is shown in **Figure 7b,** while the one using gold disc electrode is provided in **Figure 7a** for comparison. The anodic and cathodic peak difference ΔE_p is 0.069 V on the Au electrode, which is close to the theoretical value of 0.059 from Nernst equation.[18] ΔE_p is 0.34 V on the airbrushed ink electrode, suggesting a less reversible redox behavior; this value is slighter higher than screen-printed electrodes reported in the literature.[25] The change from the peak-shaped CV with a disk electrode to the sigmoidal with a microelectrode is typical and can be explained by theory.[32]

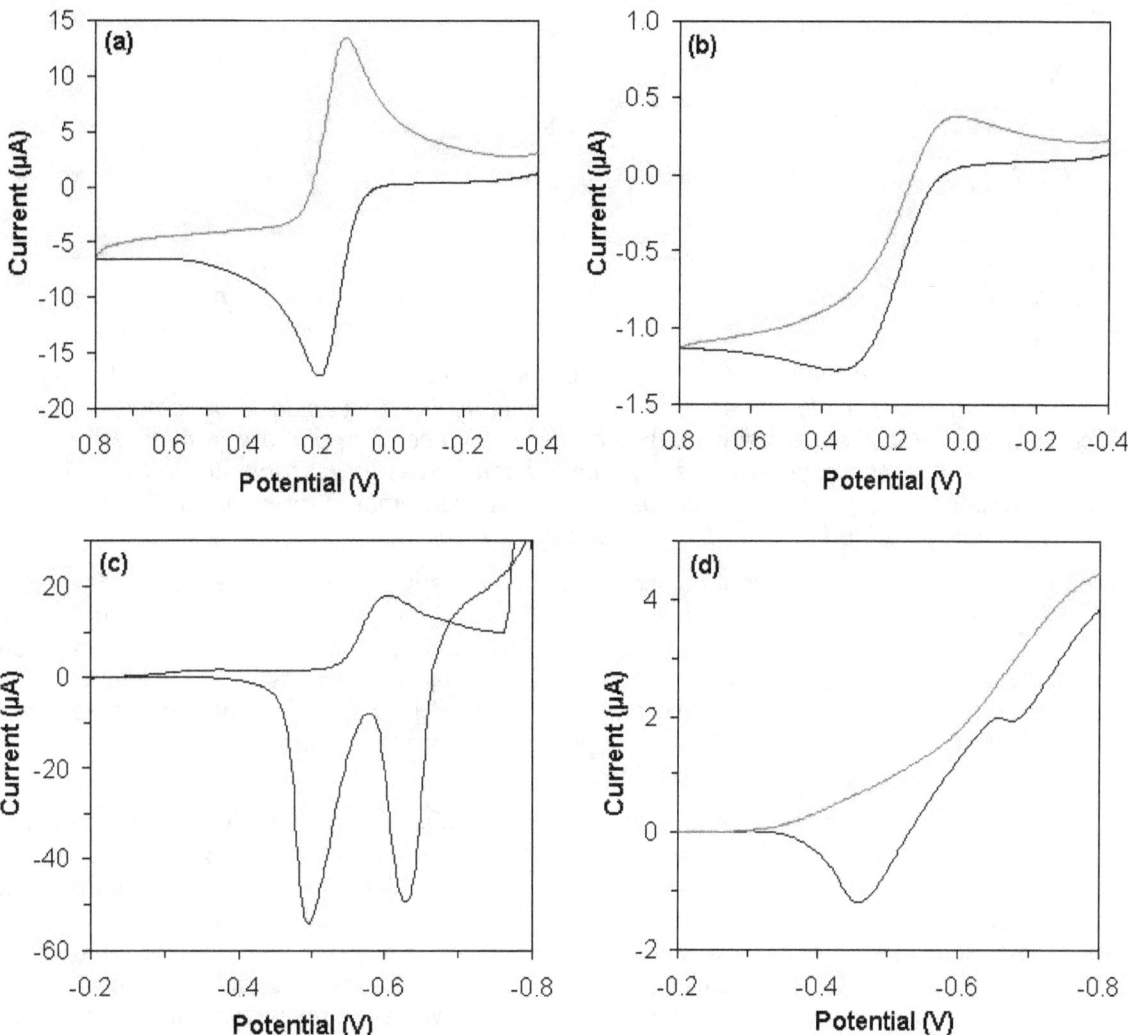

Figure 7. (a) Cyclic voltammogram of 0.005 M ferrocyanide in 0.1 M KCl using Au disc electrode. An Ag/AgCl electrode was used as the reference electrode and a platinum wire was the counter electrode. The potential was against an Ag/AgCl electrode. The potential scan rate was 0.02 V/s. (b) An airbrushed ink electrode was used. The potential scan rate was 0.005 V/s while other conditions were the same as in (a). (c) CV of 0.005 M benzyl viologen in 0.1 M KCl using Au

*electrode and the same conditions as in (a). **(d)** CV of 0.005 M benzyl viologen in 0.1 M KCl using an airbrushed ink electrode while other conditions were the same as in (b).*

Microelectrodes were also studies using benzyl viologen (BV^{2+}). **Figure 7c** shows a CV of BV^{2+} using a gold electrode while that with an airbrushed ink electrode is in **Figure 7d**. The electrochemical behavior of BV^{2+} in the electrodes is similar to those reported in the literature.[33]

Microelectrode Chronoamperometry. We also used chronoamperometry to study airbrushed ink electrodes. The resultant current-time curve from a potential step is shown in **Figure 8a.**

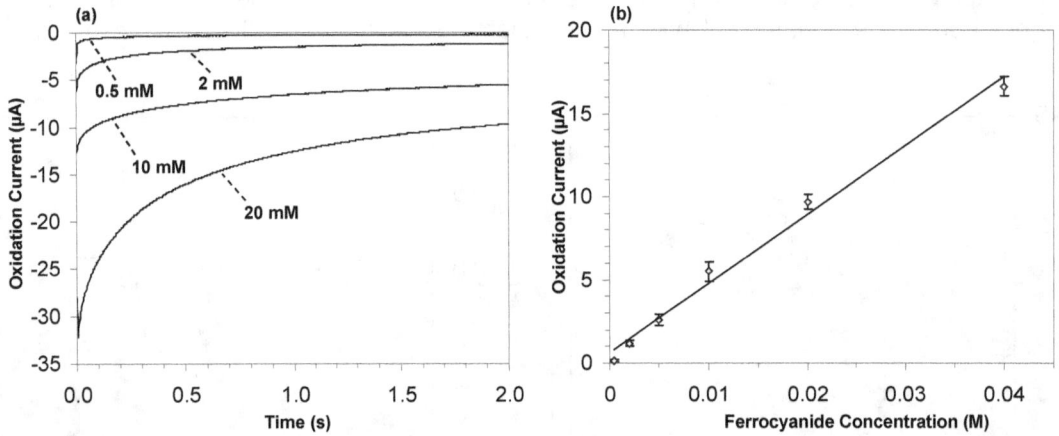

Figure 8. (a) Current-time plots obtained when chronoamperometry was performed using airbrushed ink electrodes. The concentration of ferrocyanide is indicated in the curves. An Ag/AgCl electrode was used as the reference electrode and a platinum wire was the counter electrode. (b) Resultant calibration curves for currents at the fixed time (2 s). Oxidization currents are represented by positive numbers for simplicity. The error bars indicate the standard deviations obtained from four repeat experiments.

The Cottrell equation indicates that current increases linearly with the concentration of a redox molecule at a given time. We experimentally verified this relationship by using airbrushed ink electrodes to perform chronoamperometry in ferrocyanide solutions with six different concentrations. A linear calibration curve was obtained with the regression coefficient of 0.990 when the concentration ranged from 0.5 mM to 40 mM (**Figure 8b**). Currents were selected at 2 seconds after a potential of 0.5 V is applied. The oxidation currents were corrected by subtracting the background current in 0.1 M KCl.

Ink Electrodes in a Channel. Following the procedures discussed in the *Experimental Section*, ink electrodes fabricated by airbrushing were integrated in a plastic fluidic device shown in Figure 1. These electrodes were then studied for electrochemical detection of ferrocyanide and benzyl viologen. The channels were first filled with 0.1 M KCl; electrodes were electrochemically cleaned by applying a potential of 2 V for 200 seconds. The channel was then pumped with a ferrocyanide solution. Electrochemical detection was achieved using the two-electrode configuration; the ink electrode was the working electrode while a platinum wire immersed in the well functioned as the counter and pseudo-reference electrodes.[17]

Several different concentrations of ferrocyanide were pumped through the device, resulting in different current-time curves using chronoamperometry. A calibration curve was obtained as mentioned above when an ink electrode was used in a flow cell. As shown in **Figure 9**, a linear calibration curve exists for ferrocyanide ranging from 2 mM to 20 mM.

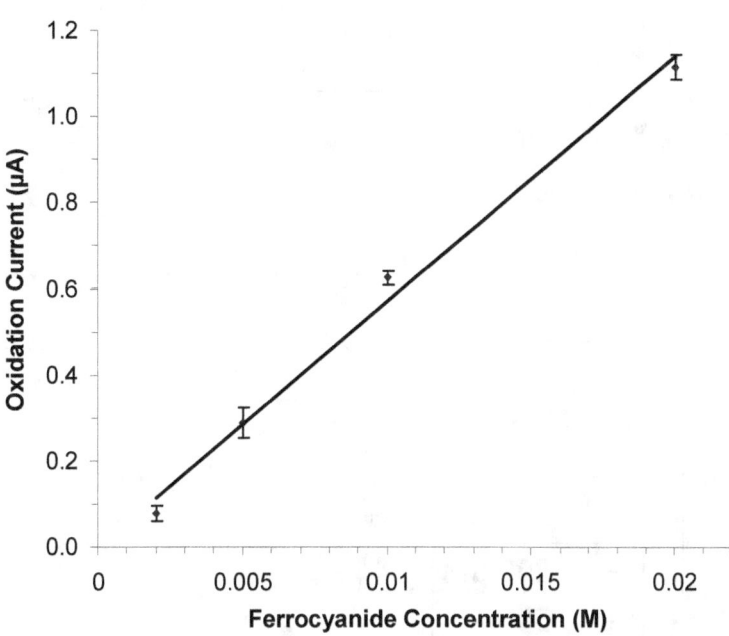

Figure 9. Calibration curves for ferrocyanide flowing in a microfluidic device. Chronoamperometric detection was achieved using an airbrushed ink electrode integrated in the device. The error bars indicate the standard deviations obtained from three repeat experiments.

Electrochemical detection of BV^{2+} using a microfabricated electrode for flow injection analysis in a microfluidic device is shown in **Figure 10**. One of four device designs in Figure 1 was used in the experiments as illustrated in Figure 10a. One inlet was filled with KCl. When it was introduced into the main channel, a background signal was obtained when a potential of 0.05 V was applied as shown in Figure 10b. When BV^{2+} in the other inlet was introduced into the main channel the current increases, indicating electrochemical detection. The signal went back to the background when the BV^{2+} flow ended. The broad BV^{2+} peak resulted from manual switching of the syringe pumps, as the plug of BV^{2+} sample injected was relatively long. Using this detection scheme, a calibration curve was obtained. The current-time curves of two concentrations of BV^{2+} were plotted, indicating the quantitative relationship between the signal and the concentration.

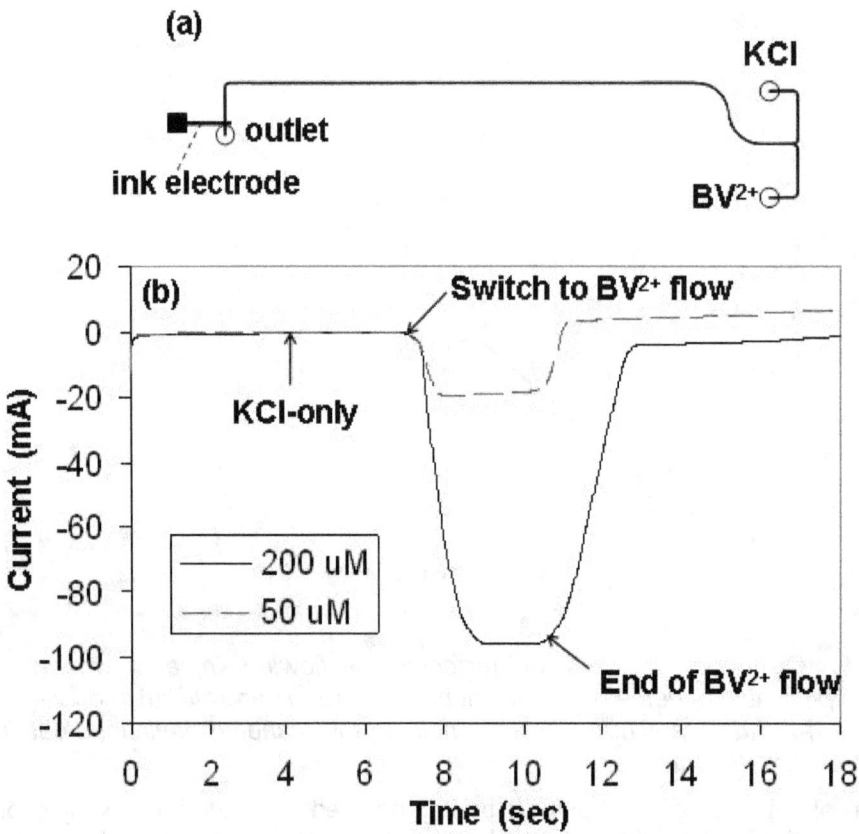

Figure 10. Chronoamperometric detection of BV^{2+} in the microfluidic device. (a) The layout of the device used (one of the designs in Figure 1). (b) The current measured as the function of time. The arrows indicate the time to introduce a KCl blank solution, a solution with BV^{2+}, and KCl again. One curve is for 50 μM BV^{2+} and the other curve is for 200 μM BV^{2+}.

Conclusions

We developed a sensing scheme to detect H_2 in gas phase by using enzyme-catalyzed electrochemical detection. The enzyme was the O_2 tolerant soluble hydrogenase of the β proteobacterium *Ralstonia eutropha*. The H_2 detection signal was generated in a flow cell in which gaseous H_2 was transported to an aqueous phase and oxidized by hydrogenase using benzyl viologen (BV^{2+}) as an artificial electron acceptor. Reduced BV^+ was subsequently detected by chronoamperometry. The net current obtained was proportional to H_2 concentration in the gas phase. Unlike popular metal catalyst-based H_2 sensors, this sensing method operates effectively at low temperatures.

The results reported here indicate that it is feasible to combine an enzymatic reaction with electrochemical detection for detecting H_2. Detection was accomplished down to 1% H_2, providing a reproducible signal with a variation of ~7% of mean signal. A linear response between current intensity and H_2 concentration was obtained up to 100% H_2. The response time of the current bench-top system includes 6 minutes of gas exposure time and 3 seconds of electrochemical detection for low concentrations of H_2. It decreases to 30-second exposure and 3-second detection for high concentrations of H_2. In addition the H_2 sensing device yielded reproducible results in the presence of 20% O_2. The limit of detection demonstrated in this work is 1% H_2, which is very relevant to the lower explosive limit of 4% H_2 in air. However, it is worthwhile noting that a lower detection limit (in ppm level) will increase the margin of safety, as illustrated by aforementioned solid-state H_2 sensors.[34] We expect the detection limit of this enzyme-based H_2 sensing method will be reduced significantly as we are moving to miniaturized devices as discussed below. Overall performance of the current system will be also improved by using enzyme preparations with higher specific activity and an improved electrolyte solution.

We also initiated the activities to adapt this enzyme-based detection scheme to a microfluidic device, because microfluidics is an ideal platform to carry out the catalytical reactions prior to electrochemical detection. In addition, we studied using airbrushing as technique for fabricating microelectrodes. Similar to screen-printing, airbrushing enabled conductive inks to be patterned to form ink electrodes. As a result, these airbrushed ink electrodes were integrated in a plastic microfluidic device to perform electrochemical detection. Both cyclic voltammetry and chronoamperometry were examined using these ink electrodes. With the potential benefits pointed out above, airbrushing could become an alternative to screen-printing, especially in situations where a commercial screen-printing machine is inaccessible.

Although we did not integrate the reference electrode in the microfluidic device, it is certainly feasible to integrate it in a device as well. A couple of approaches have been reported to integrate both working and reference electrodes in a device to achieve electrochemical detection.[35, 36] A fully integrated electrochemical detection system will lead to a completed packaged lab-on-a-chip device. In addition, other electrochemical detection methods other than chronoamperometry may be exploited to enhance the sensitivity and the detection range.

Our long-term goal is to demonstrate the function of the lab-on-a-chip device, which is to sense H_2 using enzyme-catalyzed electrochemical detection. Extremely high surface-to-volume ratio in a microfluidic device will allow for efficient interactions between the gas and aqueous phases, possibly leading to rapid reactions and high sensitivity. Efficient chemical reactions between gas and liquid samples has been demonstrated in a microfluidic system,[37] suggesting it is feasible to realize a practical miniaturized sensor based on the enzymatic reaction and electrochemical detection.

Patents, Publications, Presentations

- One provisional patent application entitled "Hydrogen Sensor Using Enzyme-Catalyzed Reaction" was filed on March 16, 2005 with application No. 60/662,504. A follow-up international PCT was filed on March 16, 2006, with application No. PCT/US2006/009495.

- Four journal papers related to this research have been published:

  o B. J. Lutz, Z. H. Fan, T. Burgdorf, and B. Friedrich, "Hydrogen sensing by enzyme-catalyzed electrochemical detection," *Anal Chem*, vol. 77, pp. 4969-75, 2005.

  o C. K. Fredrickson, Z. Xia, C. Das, R. Ferguson, F. T. Tavares, and Z. H. Fan, "Effects of Fabrication Process Parameters on the Properties of Cyclic Olefin Copolymer Microfluidic Devices," *Journal of Microelectromechanical Systems*, vol. 15, pp. 1060-1068, 2006.

  o Z. Xia, L. Cattafesta, and Z. H. Fan, "Deconvolution Microscopy for Flow Visualization in Microchannels," *Analytical Chemistry*, vol. 79, pp. 2576-2582, 2007.

  o C. Walker, Z. Xia, Z. Foster, B. J. Lutz, and Z. H. Fan, "Investigation of Airbrushing for Fabricating Microelectrodes in Microfluidic Devices," *Electroanalysis*, in press (DOI: 10.1002/elan.200704118), 2008.

- Four presentations related to this research have been made during the funding period:
    - "Fluid mixing in channels with microridges", 2007 ASME International Mechanical Engineering Congress and Exposition, Seattle, WA, November 11-15, 2007.
    - "Incorporation of Screen Printed Microelectrodes on a Microfluidic Device", Undergraduate Research Symposium, Gainesville, FL, February 17, 2007.
    - "Biology-Inspired Analysis Systems for Sensing Space-Related Species", The 57th Pittsburgh Conference, Orlando, FL, March 12-17, 2006.
    - "A Miniaturized Protein Expression Array for Detecting Toxins in Space Missions", Habitation 2006, Orlando, Florida, February 5-8, 2006.

Students from Research

- Carl K. Fredrickson, Master, Mar. 2006, Spirit AeroSystems, Inc.
- Brent Lutz, Master, Apr. 2006, Synkera Technologies Inc.
- Jackie Viren, Master, expect to graduate in 2008
- Zheng Xia, PhD, expect to graduate in 2008
- Fernando Tavares, undergraduate student, Dec. 2004, now in the graduate school of University of Michigan
- R. Ferguson, undergraduate student, May 2005, Progress Energy Inc.
- Corey Walker, undergraduate student, May 2007, now in the graduate school of University of California at Irvine
- Zachery Foster, high school student, May 2007, (now enrolled in UF)

Funding Obtained by Leveraging NASA Grant

One related proposal was funded ($70K) by National Science Foundation (NSF) with proposal No. CHE-0515711. It is entitled "Fluidic Sensors: Integrating Microfluidics with Biological Assays", covering 8/15/2005-8/14/2006. The proposal was focusing on the enzyme stability.

Collaborations

- We collaborated with Dr. Friedrich at the Humboldt University in Berlin, Germany, who provided hydrogenase for the work. We have one co-authored publication.
- We collaborated with Dr. Cattafesta at the University of Florida. We have one co-authored publication.

Acknowledgement

In addition to NASA grant, this work is supported in part by the grant of National Science Foundation (CHE-0515711) and the startup fund from the University of Florida. Drs. Mark Law and Jenshan Lin are appreciated for managing the project. We thank Florida Space Grant Consortium for the summer scholarship to Ryan Ferguson, the University of Florida for the scholarship (via the University Scholars Program) to Corey Walker, and the University of Florida Student Science Training Program to Zachery Foster.

References

(1) Turner, J. A. *Science* **2004**, *305*, 972-974.
(2) Shivaraman, M. S.; Lundstrom, I.; Svensson, C.; Hammarsten, H. *Electronics Letters* **1976**, *12*, 483-484.
(3) Lundstrom, I. *Sensors and Actuators* **1981**, *1*, 403-426.
(4) Poteat, T. L.; Lalevic, B. *Electron Device Letters* **1981**, *2*, 82-84.
(5) Hughes, R. C. *Journal of the Electrochemical Society* **1984**, *131*, C322-C322.
(6) Butler, M. A. *Journal of Applied Physics* **1985**, *58*, 2044-2050.
(7) Hunter, G.; Jefferson, G.; Madzsar, G.; Liu, C.; Wu, Q. In *Chemical Sensors II*; Butler, M., Ricco, A., Yamazoe, N., Eds.; Electrochemical Society, Inc.: Pennington, NJ, 1993, pp 256-267.
(8) Hunter, G.; Neudeck, P.; Liu, C.; Ward, B.; Wu, Q.; Dutta, P.; Frank, M.; Trimbol, J.; Fulkerson, M.; Patton, B.; Thomas, V.; Makel, D., Orlando, FL, June 12-14, 2002; IEEE International Conference on Sensors; 1126-1133.
(9) Kang, B.; Ren, F.; Gila, B.; Abernathy, C.; Peartona, S. *Applied Physics Letters* **2004**, *84*, 1123-1125.
(10) Kaltenpoth, G.; Schnabel, P.; Menke, E.; Walter, E.; Grunze, M.; Penner, R. *Anal. Chem.* **2003**, *75*, 4756-4765.
(11) Baselt, D.; Fruhberger, B.; Klaassen, E.; Cemalovic, S.; Britton, C.; Patel, S.; Mlsna, T.; McCorkle, D.; Warmack, B. *Sensors and Actuators B: Chemical* **2003**, *88*, 120-131.
(12) Sun, Y.; Tao, Z.; Chen, J.; Herricks, T.; Xia, Y. *J Am Chem Soc* **2004**, *126*, 5940-5941.
(13) Lutz, B. J.; Fan, Z. H.; Burgdorf, T.; Friedrich, B. *Anal Chem* **2005**, *77*, 4969-4975.
(14) van der Linden, E.; Faber, B. W.; Bleijlevens, B.; Burgdorf, T.; Bernhard, M.; Friedrich, B.; Albracht, S. P. *Eur J Biochem* **2004**, *271*, 801-808.
(15) Fredrickson, C. K.; Xia, Z.; Das, C.; R.Ferguson; Tavares, F. T.; Fan, Z. H. *Journal of Microelectromechanical Systems* **2006**, *15*, 1060-1068.
(16) Fox, R. W.; McDonald, A. T. *Introduction to fluid mechanics*, 5th ed.; J. Wiley: New York, 1998.
(17) Schwarz, M. A.; Galliker, B.; Fluri, K.; Kappes, T.; Hauser, P. C. *Analyst* **2001**, *126*, 147-151.
(18) Bard, A. J.; Faulkner, L. R. *Electrochemical Methods: Fundamentals and Applications*; John Wiley & Sons: New York, 1980.
(19) CH Instruments Inc. *Model 600B Electrochemical Analyzer: User's Manual*: Austin, TX, 2003.
(20) Zubay, G. *Biochemistry*, 3rd ed.; Wm. C. Brown Publishers: Dubuque, IA, 1993.
(21) Levine, I. N. *Physical Chemistry*; McGraw-Hill Book Co.: New York, 1988.
(22) Haumann, M.; Porthun, A.; Buhrke, T.; Liebisch, P.; Meyer-Klaucke, W.; Friedrich, B.; Dau, H. *Biochemistry* **2003**, *42*, 11004-11015.
(23) van der Linden, E.; Burgdorf, T.; Bernhard, M.; Bleijlevens, B.; Friedrich, B.; Albracht, S. P. *J Biol Inorg Chem* **2004**, *9*, 616-626.
(24) Turner, S.; Murrell, G. *Screen printing techniques*; Taplinger: New York, 1976.
(25) Wang, J.; Tian, B. M.; Nascimento, V. B.; Angnes, L. *Electrochimica Acta* **1998**, *43*, 3459-3465.
(26) Ball, J. C.; Scott, D. L.; Lumpp, J. K.; Daunert, S.; Wang, J.; Bachas, L. G. *Anal Chem* **2000**, *72*, 497-501.
(27) Marquette, C. A.; Lawrence, M. F.; Blum, L. J. *Anal Chem* **2006**, *78*, 959-964.
(28) Liao, W. Y.; Chou, T. C. *Anal Chem* **2006**, *78*, 4219-4223.
(29) Johirul, M.; Shiddiky, A.; Kim, R. E.; Shim, Y. B. *Electrophoresis* **2005**, *26*, 3043-3052.
(30) Tsai, D. M.; Lin, K. W.; Zen, J. M.; Chen, H. Y.; Hong, R. H. *Electrophoresis* **2005**, *26*, 3007-3012.

(31) Dember, S. *Complete airbrush techniques for commercial, technical, & industrial applications*, 1st ed.; Howard Sams and Co., Inc.: Indianapolis, 1974.
(32) Kissinger, P. T.; Heineman, W. R. *Laboratory techniques in electroanalytical chemistry*, 2nd ed.; Marcel Dekker: New York, 1996.
(33) Pang, D. W.; Abruna, H. D. *Anal Chem* **2000**, *72*, 4700-4706.
(34) Briand, D.; Wingbrant, H.; Sundgren, H.; van der Schoot, B.; Ekedahl, L. G.; Lundstrom, I.; de Rooij, N. F. *Sensors and Actuators B-Chemical* **2003**, *93*, 276-285.
(35) Ertl, P.; Emrich, C. A.; Singhal, P.; Mathies, R. A. *Anal Chem* **2004**, *76*, 3749-3755.
(36) Baldwin, R. P.; Roussel, T. J., Jr.; Crain, M. M.; Bathlagunda, V.; Jackson, D. J.; Gullapalli, J.; Conklin, J. A.; Pai, R.; Naber, J. F.; Walsh, K. M.; Keynton, R. S. *Anal Chem* **2002**, *74*, 3690-3697.
(37) Kobayashi, J.; Yuichiro, M.; Kuniaki, O.; Akiyama, R.; Ueno, M.; Kitamori, T.; Kobayashi, S. *Science* **2004**, *304*, 1305-1307.

Date: Feb. 11, 2008

6. Micro-machined Floating Element Hydrogen Flow Rate Sensor

Task PI: Dr. Mark Sheplak, Mechanical & Aerospace Engineering, University of Florida

Research Scientist: Dr. Steven Horowitz, Mechanical & Aerospace Engineering, University of Florida

Graduate Student: Tai-An Chen, Electrical & Computer Engineering, University of Florida

Research Period: August 3, 2004 to March 31, 2008

Introduction

The goal of this project is to develop a robust, miniature, silicon micromachined Moiré optical based flow rate sensor for hydrogen transport measurement applications. The miniature floating element sensor possesses optical gratings on the backside of a floating element and on the top surface of the support wafer to permit backside optical transduction. This design represents a truly flush-mounted, miniature, direct flow rate sensor that possesses immunity from electromagnetic interference (EMI) and transverse element movement due to pressure fluctuations and/or vibrations. The flow disturbance is minimal because the incident and reflected light comes through the backside of the Pyrex wafer and the floating element and associated gap are covered with a thin polymer coating to eliminate gap-roughness effects. In addition, there is no thermal or electrical energy transfer from the sensor to the hydrogen medium.

Background

Technical Approach

MEMS-Optical Floating Element Flow Rate Sensor:

The sensor consists of a silicon floating element and Pyrex support structure with optical gratings patterned on both structures to allow backside optical transduction (Figure 1A). The sensor functions by measuring the integrated wall shear stress generated by the flow rate, τ, acting on the floating element, which in turn deflects the tethers. Each set of aluminum optical gratings have a slightly different pitch to realize that results in a Moiré pattern (Figure 1B). The overall spatial period of the Moiré pattern, G, is determined from the pitch of the gratings on the floating element, g_1 and the pitch of the gratings on the support structure g_2

$$\frac{1}{g_1} - \frac{1}{g_2} = \frac{1}{G} \qquad (1)$$

The Moiré amplification over the mechanical displacement is determined by the ratio

$$G/g_2 \qquad (2)$$

A protective coating will be applied to the surface of the shear stress sensor to protect the cavity between the floating element and the support structure of the sensor from the environment.

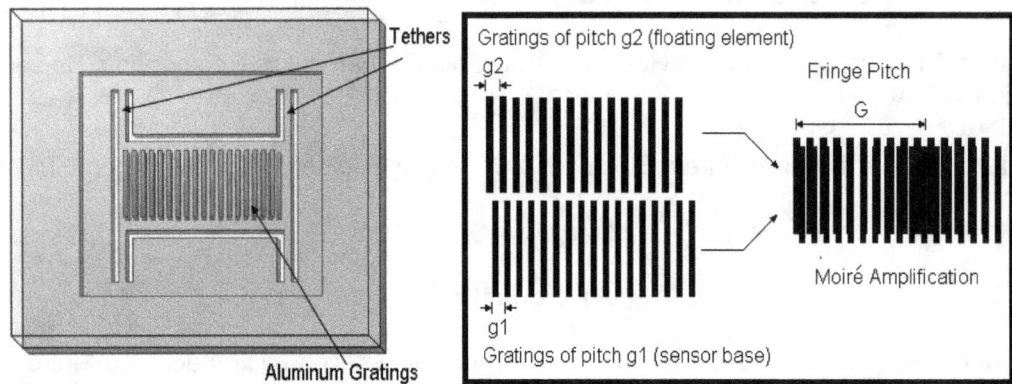

Figure 11: Optical Moiré based flow rate sensing concept. A) Top view schematic of sensor B) Graphical illustration of Moiré amplification arising from two slightly different gratings.

Several key benefits using the optical Moiré detection scheme are as follows:

- The floating element flow sensor uses optoelectronics far removed from the measurement site such that electromagnetic interference (EMI) does not affect the working sensor.
- The Moiré amplification increases the sensitivity in detecting mechanical motion.
- Due to the amplification, slight displacements in the mechanical structure greatly shift the Moiré pattern, allowing for minute motion detection.
- Since the Moiré scheme uses phase detection, fluctuations in the intensity do not affect the phase difference.

Sensor Design

Several sensor specifications associated with various candidate flows were targeted as listed in Table 1 where τ_{max} is the maximum shear stress to be measured, f_{res} is the targeted resonant frequency to provide adequate temporal resolution, L_e, W_e and T_e are the designed floating element length, width and thickness, and L_t, W_t and T_t are the designed tether length, width and thickness, respectively. The expected sensitivity and maximum deflection, ΔL_{max}, are also shown, along with the deflection, ΔL, for a 5 Pa input.

Table 4. Candidate flow sensor specifications for several applications.

Target	L_t [um]	W_t [um]	T_t [um]	L_e [um]	W_e [um]	T_e [um]	f_{res} [kHz]	Sens [nm/Pa]	ΔL_{max} [nm]	τ_{max} [Pa]	ΔL @ 5Pa [nm]
5 Pa/2kHz	1250	10	45	1500	1000	10	2.3	98.5	2167	22	492.5
	1250	10	45	1500	1000	45	1.1	98.5	2167	22	492.5
5 Pa/5kHz	1100	20	45	2500	1500	10	5	20.8	4279	205	104.5
	1100	20	45	2500	1500	45	2.4	20.8	4279	205	104.5
50 Pa/5kHz	800	20	45	1500	1500	10	10.1	5.15	4276	830	25.76
	800	20	45	1500	1500	45	4.8	5.15	4276	830	25.76
5 Pa/1kHz	1400	10	45	1500	1800	10	1.2	247	2228	9	1237
	1400	10	45	1500	1800	45	0.7	247	2228	9	1237

Experimental

The current optical detection setup consists primarily of an optical fiber bundle array and a 16 element photodiode array used for the optical transduction detection. The shear stress sensor is flush mounted on an optical base with the floating element exposed on in the flow field and the backside exposed to the optical sensing scheme. A source laser beam, with a cross sectional area of 4 mm × 1 mm, is mounted on an adjustable optical package to illuminate the Moiré gratings on the shear stress sensor.

Due to the size differences between the length of the Moiré pattern and the length of the 16 element photodiode array, the optical fiber bundle (Figure 2) is designed to map these features accordingly. The fiber bundle array contains 16 individual fibers each with a diameter of 62.5 μm. The fiber plug on the sensor side of the fiber bundle designed such that the total length of the sensor side is 1 mm and the fiber plug on the photodetector side is 16 mm. The fan-out inside the fiber bundle array essentially acts like an optical magnification of 16× to normalize the Moiré pattern across the photodiode array.

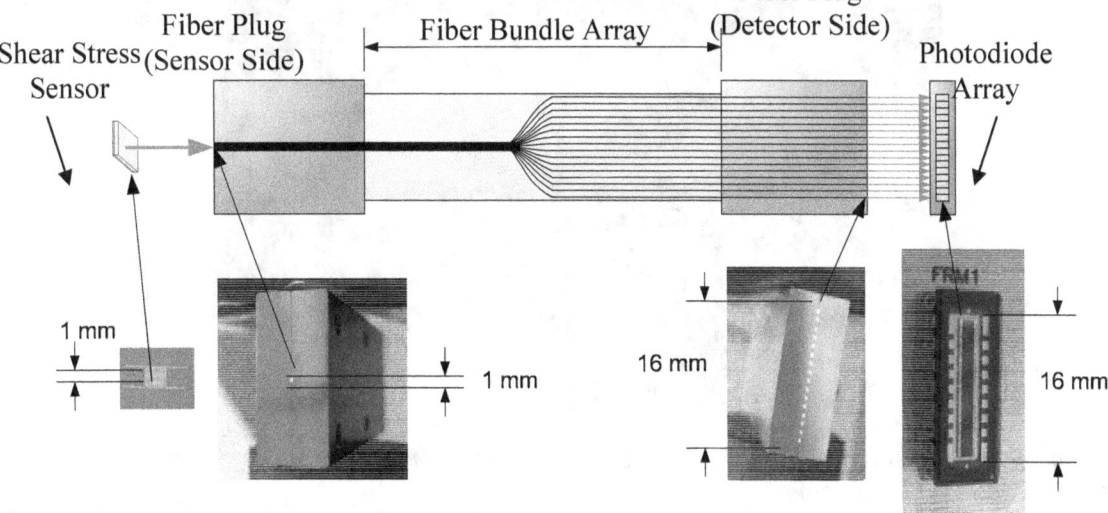

Figure 2: Schematic and photographs of the optical fiber bundle in relation to the flow sensor and optical detector to illustrate the mapping of the Moiré pattern to the photodiode array.

A pre-amplifier circuit is designed to adjust the gain of the signal received from the photodetectors as well as to reduce the noise levels before sending this information to a data acquisition system. A National Instruments DAQ system is used to simultaneously record all 16 channels from the photodiode array. The data is post processed using a curvefit algorithm to extract the Moiré period and compared to known shear stress values for sensor calibration. The overall optical setup is illustrated in the following diagram (Figure 3) and photograph (Figure 4).

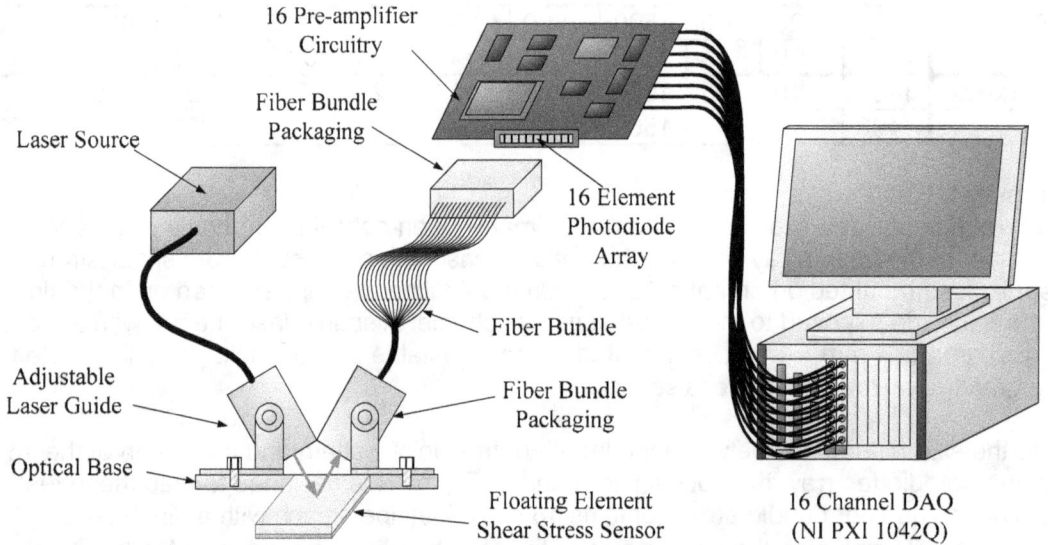

Figure 3: Schematic of the optical setup for obtaining the Moiré patterns from the flow sensor to the data acquisition system.

Figure 4: Photograph of the optical setup for obtaining the Moiré patterns from the flow sensor to the data acquisition system.

Static Calibration

For the static calibration, a 2-D laminar flow cell (Figure 5) will be used to provide a mean shear flow for the sensor. The semi-infinite parallel plate flow cell will provide a pressure driven Poiseuille flow and the wall shear stress can be calculated by:

$$\tau = \frac{h}{2}\frac{dP}{dx}$$

where h is the height of the flow cell, dP is the differential pressure between two pressure taps and dx is the distance between the two pressure tap measurements.

Experimental Setup for Static Calibration

The channel length of the flow cell is 330.2 mm and the cross section dimensions of the flow cell are 100 mm x 1 mm. The pressure taps are located along the path of the flow and pressure measurements will be observed used a Heise Pressure meter. The sensor will be flush mounted in a packaging plug and attached to the flow cell.

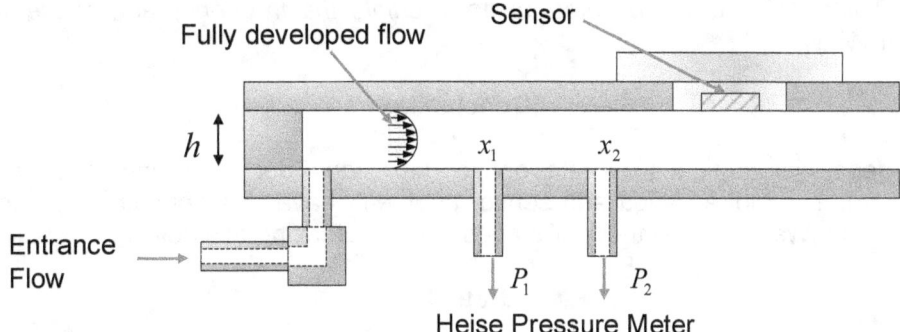

Figure 5: Laminar flow cell providing a pressure driven Poiseuille Flow.

Results and Discussion

Currently an optical testbed using two Pyrex plates containing aluminum gratings of various pitches has been set up to test the optics. A mechanical nanopostioner is used to mechanically shift one of the grating plates to simulate the effect of the Moiré pattern. A real-time curve-fitting algorithm has been implemented using LabVIEW to detect the Moiré frequency and phase information (Figure 6).

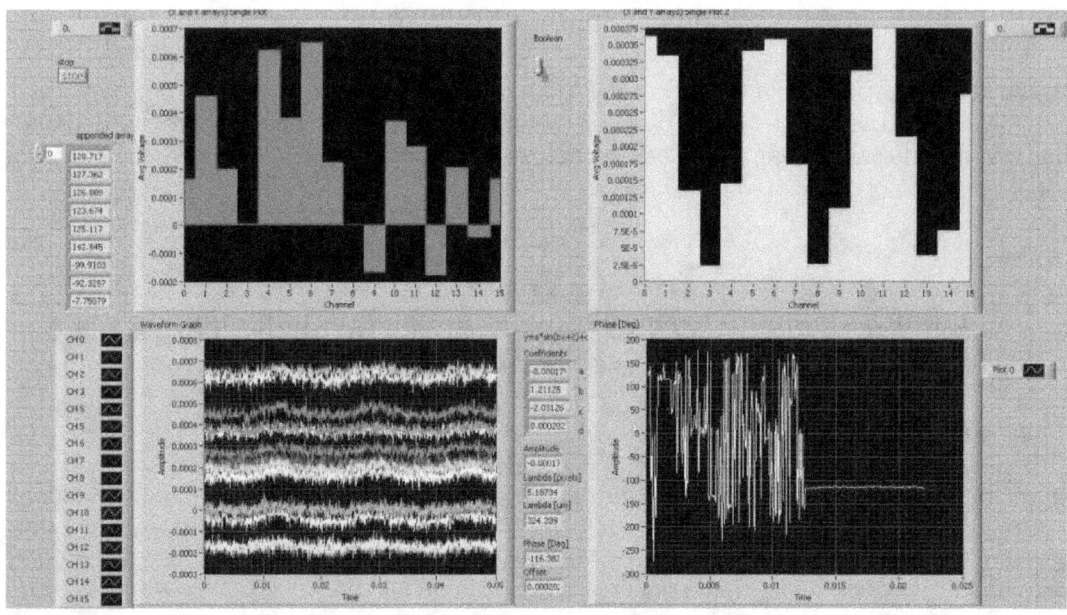

Figure 6. Optical Moiré detection program to calculate the frequency and phase information using LabVIEW.

Conclusion

A second generation optical flow rate sensor has been designed and is currently being microfabricated. The Moiré period and sensor geometry have been optimized to increase the overall sensitivity. We are still in the final stages of characterizing this device.

Patents

None

Publications

Horowitz, S., Chen, T., Cattafesta, L., Sheplak, M., Nishida, T., and Chandrasekaran, V., "Optical Flow Sensor Using Geometric Moire Interferometry," SAE Paper 2004-01-2267, 34th International Conference on Environmental Systems Colorado Springs, CO, 2004.

Sheplak, M., Cattafesta, L., and Tian, Y., "Micromachined Shear Stress Sensors for Flow Control Applications," IUTAM Symposium on Flow Control and MEMS, Springer, ed. J. F. Morrison, pp. 67-73, 2006.

Students from Research

Steve Horowitz: PhD Student.

Funding Received by Leveraging the NASA grant

"Moiré-Based Optical MEMS Shear Stress Sensor Technology," Office of Naval Research

7. Ultra-wideband Communication for Tiny Low Power Radios

Task PI: Dr. Kenneth K. O, Electrical & Computer Engineering

A final report was submitted for this project. It is provided again.

Research Period: August 3, 2004 to April 30, 2006

Abstract
A wireless network utilizing one way communication from distributed sensors to a central monitoring station or a more central base station with more processing and communication capabilities is proposed for H_2 sensing applications. The communication link utilizes pulse position modulated ultra-wide band signals converted up to ~5.8 GHz. A Schottky diode with cut-off frequency greater than 1 THz has been realized in foundry CMOS and utilized to demonstrate an ultra-wide band amplitude modulation detector. An on-chip antenna for 5.8-GHz operation, which is suitable for communication to a base station located more than 100 meters has been demonstrated. The antenna is ~6 mm x 0.2 mm in size. Integration of the antenna will lower cost of sensor nodes.

Project Goals
The goals are to develop radio architecture for communication up to 10 m which enables radical reduction of power consumption, and to demonstrate the concept in a single chip RF transceiver (~3 mm x 1.5 mm) with integrated antennas. Anticipated results are ultra wide band (UWB) radio architectures with reduced power consumption which can be implemented as an RF system on a chip, and a single chip RF transceiver with integrated antennas for UWB communication.

Background
A wireless network utilizing one way communication from distributed sensors to a central monitoring station or a more central base station with more processing and communication capabilities is proposed for H_2 sensing applications. For the lifetime of 10 years without any maintenance (including power source changes) desired by NASA, low power consumption is must. Because of the potentially large number, the sensors should be low cost. For improved flexibility of use, they should also be small. Often, power consumption of radio links dominates power consumption of sensor networks. By lowering the power consumption of radio links, the operation time of distributed H_2 sensors between battery recharges or replacements can be increased. Lowering power consumption will also reduce battery size, thus lowering the cost and size of sensor nodes.

Accomplishments
6-mm long linear monopoles with a sleeve (Fig. 1(a)) have been fabricated on 20-Ω-cm silicon substrate with a 3-μm thick SiO_2 layer. The sleeve length is ~600 μm. The antenna pair gain (G_a) defined in Fig. 1(b) has been characterized at 5.8 GHz versus separation. The pair was located at 52 cm from the ground. Assuming that Ga of -97dB is acceptable to realize communication link, these antennas are sufficient for communication up to 40 m. Compared to the 3-mm long zigzag pair operating at 24 GHz, the range has been increased by around 3X. This is indeed significant improvement.

Based on the measured antenna gain of ~-100 dB between the zigzag dipole antenna and a horn with 20 dBi separated by 90m, these monopole antennas should be adequate for communication up to 100 m. To verify this, a 5.8-GHz patch antenna with 11 dBi gain (Fig. 2) for a base station has been purchased. A 0-dBm sine wave was transmitted using the 6-mm monopole antenna and was picked it up with the patch antenna located 310 m away. This exceeds the original target of 100 m by 3X.

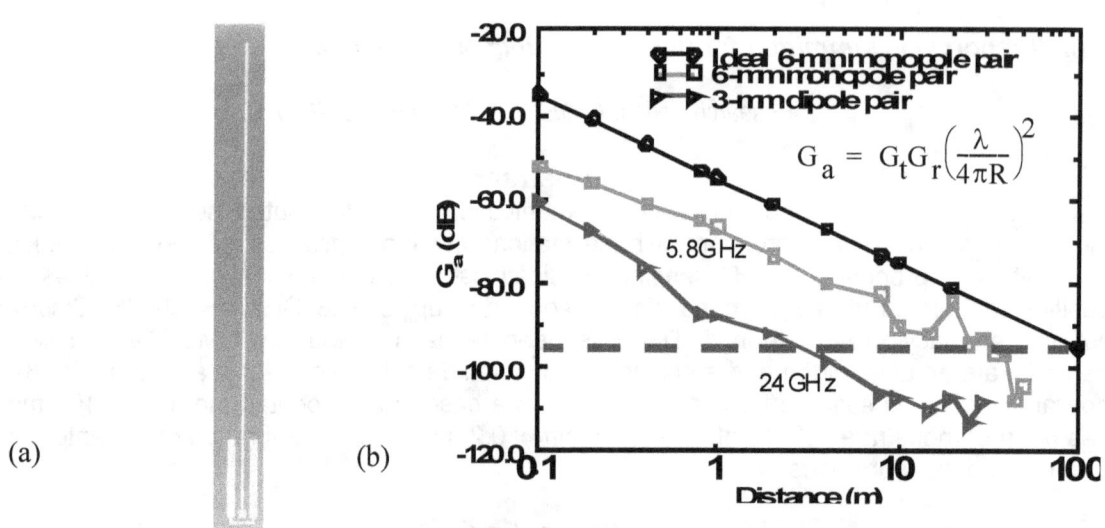

Figure 1(a). A micro-photograph of the monopole antenna, (b) plots of antenna pair gain, G_a. The antenna pair was located 52 cm above the ground.

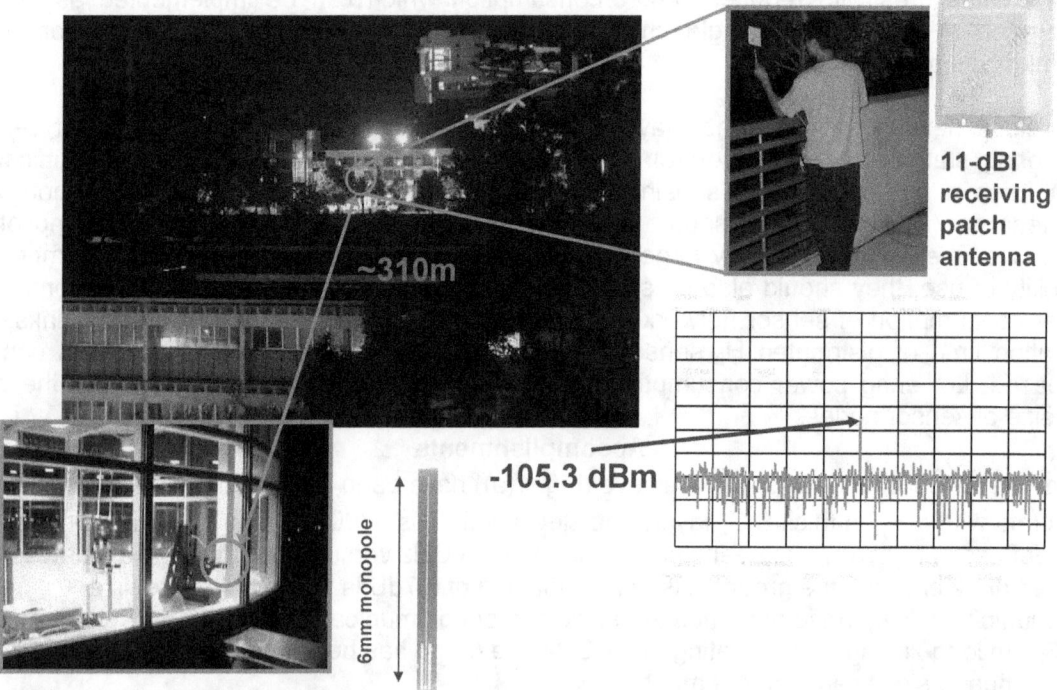

Figure 2. A 6-mm on-chip monopole was used to transmit a 0- dBm 5.8- GHz sine wave and a patch antenna located 300 m away was used to pick up the wave.

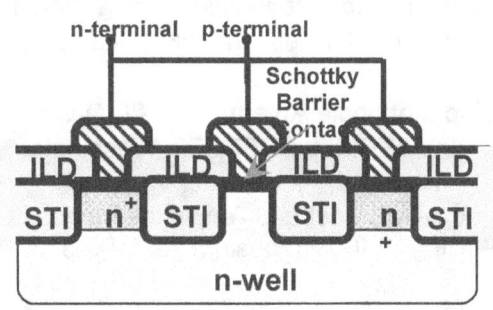 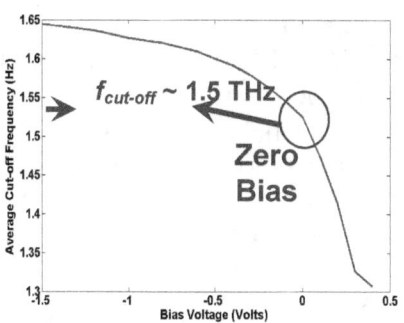

Figure 3. A cross section of Schottky diodes fabricated in 130-nm CMOS. The cut-off frequency is ~1.5 THz.

The support was also used to finish off the Schottky diode work started using the 2003 funding. The diode achieves cut-off frequency of 1.5 THz. This is the highest ever reported for structures fabricated using mainstream silicon technologies. This diode is also used to demonstrate 182-GHz demodulator using low cost 130-nm CMOS technology. This is a ground breaking work indicating that THz applications can be addressed using CMOS technology used to fabricate microprocessors and memory circuits. Three white papers have been submitted to support a full blown research on THz CMOS circuits.

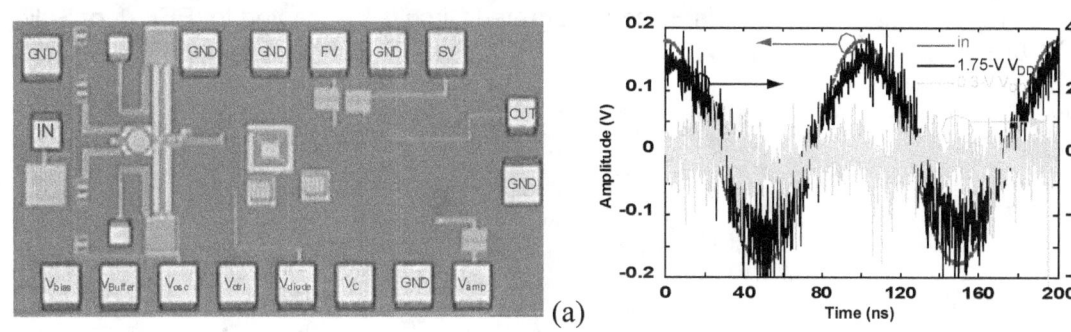

Figure 4, (a) 182-GHz modulator and demodulator implemented using 130-nm CMOS technology. (b) The input and output waveforms of the modulator/demodulator.

Publications

[1] S. Sankaran, and K. K. O, "Schottky Barrier Diodes for Millimeter Wave and Detection in a Foundry CMOS Process," IEEE Electron Device Letters, vol. 26, no. 7, pp. 492-494, July, 2005.
[2] S. Sankaran, and K. K. O, "A Schottky Diode with Cut-off Frequency of 400 GHz Fabricated in 0.18-μm CMOS," Electronics Letters vol. 41, no. 8, pp. 506-508, Apr. 2005.
[3] K. K. O, K. Kim, B. A. Floyd, J. Mehta, H. Yoon, C.-M. Hung, D. Bravo, T. Dickson, X. Guo, R. Li, N. Trichy, J. Caserta, W. Bomstad, J. Branch, D.-J. Yang, J. Bohorquez, E. Seok, L. Gao, A.Sugavanam, J.-J. Lin, J. Chen, and J. Brewer, "On-chip Antennas in Silicon Integrated Circuits and Their Applications," **(Invited)** IEEE Trans. on Electron Devices, vol. 52, no. 7, pp. 1312-1323, July 2005.
[4] K. K. O, K. Kim, B. Floyd, J. Mehta, H. Yoon, C.-M. Hung, D. Bravo, T. Dickson, X. Guo, R. Li, N. Trichy, J. Caserta, W. Bomstad, J. Branch, D.-J. Yang, J. Bohorquez, J. Chen, E.-Y. Seok, L. Gao, A. Sugavanam, J.-J. Lin, S. Yu, C. Cao, M.-H. Hwang, Y.-P. Ding, S.-H. Hwang, H. Wu,

N. Zhang, and J. E. Brewer, "The Feasibility of On-Chip Interconnection using Antennas," IEEE/ACM International Conference on Computer Aided Design, pp. 979-984, San Jose, CA, Nov. 2005.

[5] J.-J. Lin, H.-T. Wu, and K. K. O, "Compact On-Chip Monopole Antennas on 20-Ω-cm Silicon Substrates for Operation in the 5.8-GHz ISM Band," 2005 IEDM, pp. 967-970, Washington, DC, Dec. 2005.

[6] S. Sankaran and K. K. O, "A Schottky Barrier Diode Ultra-Wideband Amplitude Modulation (AM) Detector in Foundry CMOS Technology," 2006 IEEE RFIC Symposium, pp. 309-312, June 2006, San Francisco, CA.

[7] E. Seok, C. Cao, S. Sankaran, and K. K. O, "A Millimeter-Wave Schottky Diode Detector in 130-nm CMOS Technology," 2006 Symposium on VLSI Circuits, pp. 178-179, June 2006, Honolulu, HI.

[8] K. K. O, K. Kim, B. Floyd, J. Mehta, H. Yoon, C.-M. Hung, D. Bravo, T. Dickson, X. Guo, R. Li, N. Trichy, J. Caserta, W. Bomstad, J. Branch, D.-J. Yang, J. Bohorquez, J. Chen, E.-Y. Seok, J. E. Brewer, L. Gao, A. Sugavanam, J.-J. Lin[*], Y. Su, C. Cao[*], M.-H. Hwang, Y.-P. Ding, Z. Li, S.-H. Hwang, H. Wu, S. Sankaran, and N. Zhang, "Silicon Integrated Circuits Incorporating Antennas," **(Invited)** 2006 IEEE Custom Integrated Circuits Conference, pp. 473-480, Sep. 2006, San Jose, CA.

[9] S. Sankaran and K. K. O, "A Ultra-Wideband Amplitude Modulation (AM) Detector using Schottky Barrier Diodes Fabricated in Foundry CMOS Technology," IEEE J. of Solid-State Circuits, vol. 42, no. 5, pp. 1058-1064, May, 2007.

[10] J.-J. Lin, H.-T. Wu, S. Yu, L. Gao, A. Sugavanam, J. E. Brewer, and K. K. O, "Communication Using Antennas Fabricated in Silicon Integrated Circuits," Accepted to IEEE J. of Solid-State Circuits.

www.ingramcontent.com/pod-product-compliance
Lightning Source LLC
Chambersburg PA
CBHW081731170526
45167CB00009B/3782